MW01519817

NATURE'S YEAR

NATURE'S YEAR

Changing Seasons in Central and Eastern Ontario

Drew Monkman

DUNDURN
TORONTO

Copyright © Drew Monkman, 2012

All rights reserved. No part of this publication may be reproduced, stored in a retrieval system, or transmitted in any form or by any means, electronic, mechanical, photocopying, recording, or otherwise (except for brief passages for purposes of review) without the prior permission of Dundurn Press. Permission to photocopy should be requested from Access Copyright.

Editors: Jennifer McKnight and Allison Hirst
Design: Jesse Hooper
Printer: Webcom

Library and Archives Canada Cataloguing in Publication

Monkman, Drew, 1952-
    Nature's year : changing seasons in central and eastern Ontario / Drew Monkman.

Includes bibliographical references and index.
Issued also in electronic formats.
ISBN 978-1-4597-0183-0

    1. Natural history--Ontario--Guidebooks. 2. Seasons--Ontario--Guidebooks. 3. Ontario--Guidebooks. I. Title.

QH106.2.O5M64 2012   508.713   C2011-907990-9

1   2   3   4   5   16   15   14   13   12

We acknowledge the support of the **Canada Council for the Arts** and the **Ontario Arts Council** for our publishing program. We also acknowledge the financial support of the **Government of Canada** through the **Canada Book Fund** and **Livres Canada Books**, and the **Government of Ontario** through the **Ontario Book Publishing Tax Credit** and the **Ontario Media Development Corporation**.

Front cover images (top to bottom): Fall foliage, Karl Egressy; winter tamaracks, Terry Carpenter; pastel greens of spring trees, Terry Carpenter; summer stream, Terry Carpenter.

Care has been taken to trace the ownership of copyright material used in this book. The author and the publisher welcome any information enabling them to rectify any references or credits in subsequent editions.

J. Kirk Howard, President

Printed and bound in Canada.
www.dundurn.com

Dundurn	Gazelle Book Services Limited	Dundurn
3 Church Street, Suite 500	White Cross Mills	2250 Military Road
Toronto, Ontario, Canada	High Town, Lancaster, England	Tonawanda, NY
M5E 1M2	LA1 4XS	U.S.A. 14150

MIX
Paper from
responsible sources
FSC® C004071

To my granddaughter, Anouk, who inspires me to care more deeply than ever about the future of the natural world

CONTENTS

ACKNOWLEDGEMENTS

A BOOK OF THIS SORT WOULD NEVER HAVE BEEN POSSIBLE WITHOUT THE ASSISTANCE of a great many people. I am especially grateful, however, to the late Doug Sadler, a Peterborough naturalist, writer, and educator, who in many ways was my mentor and who helped me to look at the natural world in a way that filled me with wonder and tantalizing questions to ponder.

I also wish to extend my sincere thanks to Kim Caldwell and Jean-Paul Efford, whose superb drawings illustrate the chapters of this book. I am also very grateful to have been able to use the excellent photographs of Terry Carpenter, Harold Clark (author photo), Joe Crowley, Karl Egressy, Matthew Garvin, Rick Stankiewicz, Tyler Wheeldon, and the Ontario Ministry of Natural Resources. Thanks also are extended to Allison Hirst, senior editor at Dundurn, for her expertise and patience in editing the manuscript, and to Jeff Macklin, a graphic designer at Jackson Creek Press, for all his hard work in preparing the maps and bird abundance charts.

The accuracy and completeness of the book was also greatly enhanced by the following individuals who gave generously of their time to review sections of the manuscript and to share their expertise. They include a number of people from Trent University, Parks Canada, the Natural Heritage Information Centre (NHIC), and the Ontario Ministry of Natural Resources (OMNR).

Thanks are extended to Peter Adams (Trent University), Jerry Ball, David Beresford (Trent University), Bob Bergmann (OMNR), Tony Bigg, David Bland (Parks Canada), John Bottomley, Peter Burke, Bill Crins (OMNR), Paul Elliott (Trent University), Jason Hollinger, Steve Kerr (OMNR), Peter Lafleur (Trent University), Janine McLeod (Alderville First Nation Black Oak Savanna), Mike McMurtry (NHIC), Brian Morin (Parks Canada), Erica Nol (Trent University), Mike Oldham (NHIC), Kathy Parker, Martin Parker, Chris Risley (OMNR), Rick Rosatte (Trent University/OMNR), Rick Stankiewicz, Don Sutherland (NHIC), and Brad White (Trent University).

While many have been supportive of this work—and no one more so than my wife, Michelle—the responsibility for any errors remains with me. Any such that are reported to the publisher or I will be rectified in subsequent editions.

To be interested in the changing seasons is a happier state of mind than to be hopelessly in love with spring.

— *George Santayana*

INTRODUCTION

ONE OF THE GREATEST PLEASURES TO BE DERIVED FROM THE REGULAR OBSERVATION of the natural world is to witness the passage of time as one season slowly slips into the next. The predictability of seasonal change comes from the recurrence every spring, summer, fall, and winter of natural events that often take place within a week or so of the same date each year. Such events provide a reassuring counterbalance to the onslaught of change and uncertainty that characterizes modern life. In nature we can rest assured that the barely perceptible lengthening of a January day is leading to another spring, that April mornings will ring with birdsong, and that maples will glow in yellows, oranges, and reds come the end of September.

Nature's Year is a month-by-month almanac of key events occurring in the natural world over the course of one year. The sequence and general timing of events applies to all of central and eastern Ontario, a region extending from the Bruce Peninsula and Georgian Bay in the west to Ottawa and Cornwall in the east. Much of this region is also known as "cottage country," a term that refers to popular cottaging and vacation areas such as Parry Sound, Muskoka, the Kawarthas, the Haliburton Highlands, Land O' Lakes, and the Rideau Lakes.

Like a celestial clock, events in nature tick off the time of the season. They help us to become more aware of the continuity of seasonal progression and to find and identify species of interest. To a large extent, "seeing" means knowing what to expect. Awareness that Bald Eagles are a real possibility over a half-frozen December lake greatly facilitates actually seeing them. In fact, the cornerstone of most plant and animal identification is the knowledge of what to anticipate given the time of year and type of habitat. A seasons-based approach to natural history, therefore, provides a way to mentally organize and retrieve information about plants and animals. Otherwise, much of the content of field guides and other nature books can seem like an impenetrable mass of information.

My purpose in writing this book is also to help people pay more attention to the "near-at-hand" natural world. Thanks to modern media, we are forever hearing about the big picture of nature — from the amazing biodiversity of the rainforest and the mammals of the plains of Africa to the effects of El Niño and climate change. But how many of us ever pay full attention to what's going on in our own backyard, neighbourhood, or cottage community? My hope is that *Nature's Year* will provide readers with an intimate record of "nearby" nature over the course of a single year.

The events that constitute nature's annual cycle are practically limitless. In central and eastern Ontario, there are approximately 60 species of mammals, 17 amphibians, 16 snakes, 8 turtles, 8 salamanders, 10 frogs and toads, 100 or more butterflies, at least 135 dragonflies and damselflies, and over 200 breeding birds. At least another 100 bird species pass through the area annually but do not stop to breed. As for vascular plants, the number of species is probably in the range of 1,800 to 2,000.[1]

Each of these animal and plant species responds to the changing seasons through a series of datable events, be it emergence from hibernation or the emergence of the first leaves. No one book could ever cover all of these events for every species — if, in fact, they were all even documented. The events that I have chosen to include simply provide a glimpse into the lives of *some* species, given the time of year. The dates used in this book are based on many years of observations, both by me and by other naturalists across central and eastern Ontario with whom I consulted. Some dates, however, are "best estimates." For example, while arrival and departure dates for nearly all bird species have been carefully documented for decades, the dates for some mammal and insect activities are not known as accurately.

The science of observing and recording the annual cycle of first events — be it spring's first Tree Swallow, lilac bloom, or spring peeper song — is known as *phenology*. Keeping track of the dates of first happenings from year to year not only enhances the pleasure of seeking out these events but provides a measure of order and predictability to phenomena in nature. It also serves as a way of being attentive to all that surrounds us. Phenology helps us to see the land as a whole. It is interesting to compare, for any given occurrence, what other events are happening at around the same time. For example, when spring peepers are calling in late April, American Woodcock are displaying, Walleye are spawning, American elms are in flower, the ice has probably just gone out of the lakes, the sun sets shortly after 8:00 p.m., and Orion is low in the western sky.

Depending on the species, the date of "first occurrence" is not always the same each year. Factors such as abnormal weather patterns or the influence of an El Niño can delay or hasten events. Climate change, in particular, is having an effect on phenological dates, especially with respect to those events occurring in the early spring. As a general rule, the year-to-year variability of events decreases as the spring advances. For example, there is much more variability in the arrival dates of early spring birds as compared to those birds that arrive in May. Also, by its very nature, a first event means that the event in question may not yet be widespread or easy to observe; when a given species of bird returns in the spring, there are usually only a small number of individuals for the first week or so, with larger numbers arriving later. As an example, although the first Red-eyed Vireos arrive in mid-May, the bird is not common until late in the month.

American biologist and writer Bernd Heinrich suggests that "most of us are like sleepwalkers here, because we notice so little."[2] For so many people, nature has been reduced to little more than pretty landscapes and the green blur that rushes past on the other side of closed car windows. I hope that this book will help people to become more aware of all that surrounds us in central and eastern Ontario and in this way develop both a greater sense of place and sense of season. If we are ever going to take care of this planet, we have to start by knowing and caring about our own region. Human nature is such that in the end we will protect only what we love and love only what we know. The knowing, however, must go beyond simply putting names to what we find in nature. We need to understand the natural world as a dynamic entity of countless interrelationships, the complexity of which we can only begin to understand. The knowing also needs to be done in a context of place — in this case central and eastern Ontario — and in a context of time — the changing seasons.

WHY DO WE HAVE SEASONS?

Let's begin by thinking of a globe. You have no doubt noticed that a globe is tilted, as is the Earth. In other words, the imaginary line between the Earth's north and south poles is not vertical but on a 23.5 degree angle. Because the Earth is also rotating on its own axis, this means that the northern hemisphere ends up being tilted toward the sun for part of the year — our spring and summer — and away from the sun for part of the year — our fall and winter. The main consequence of this tilting is a huge difference in the amount of heating of the Earth's surface that occurs from one season to the next. In summer, sunlight strikes our part of the globe much more perpendicularly than at other times of the year and therefore heats the Earth much more efficiently. The solar radiation also takes a somewhat shorter path through the energy-absorbing atmosphere before striking the Earth. In winter, on the other hand, the sun casts a weaker, angled light from its position much lower in the southern sky. The sunlight must also travel through more atmosphere. Therefore, far less heating of the Earth's surface occurs. The difference in heating between the summer and winter can also be illustrated by pointing a flashlight at a tabletop. Summer is akin to shining the beam directly down on the table from straight above so that the light focuses on a small area. The table top will soon feel warm to the touch. For winter conditions, angle the beam to the side so that the light scatters over a larger area. Far less heating occurs.

The significance of the difference in heating between summer and winter is profound. All life responds, be it the phenomenal plant growth and birdsong of solar radiation-rich June or the plant dormancy and avian silence resulting from the weak sunlight of December. The tilted axis is also the reason why there are more hours of daylight in the summer and fewer hours of daylight in the winter. This, too, makes a huge difference in the lives of plants and animals. One of the main reasons that birds migrate north in the spring — instead of staying in the tropics, for example — is to take advantage of the longer days of the temperate zone summer. The longer days mean more time for birds to gather food to feed their young, hence greater reproductive success.

ORGANIZATION OF THE BOOK

In order to best meet the reader's needs, interests, and available time, the natural events included for each month are presented at different levels of detail. A brief initial essay attempts to capture the mood of the month and to introduce some of the most distinctive natural events taking place. Each of the eight areas of interest — birds, mammals, amphibians and reptiles, fish, insects and other invertebrates, plants and fungi, weather, and the night sky — are then examined in detail, starting with a list of key events and the approximate time of month each event usually first occurs. Some events may occur all month long, while others are more specifically associated with the early month (from the 1st to the 10th), mid-month (from the 11th to the 20th) or late month (from the 21st to the 31st).

Some of these events (those followed by "See page xx") are then described in more detail. Various tables are also used to summarize information. A chart showing the seasonal abundance of common bird species is located in Appendix 1. When referring to the book to find information on a given event in nature that you may notice, you should also check the month immediately before and/or after.

THE SETTING

For the purposes of *Nature's Year*, central and eastern Ontario covers the geographical area extending from the Bruce Peninsula and Georgian Bay in the west to Ottawa and Cornwall in the east. It also includes the sub-region of eastern Ontario, the wedge-shaped area between the Ottawa and St. Lawrence Rivers. The book covers nearly all of so-called cottage country (Parry Sound, Muskoka, Haliburton, Bancroft, Kawartha Lakes, Land O' Lakes, and Rideau Lakes) and includes the counties of Grey, Bruce, Huron, Wellington, Dufferin, York, Simcoe, Durham, Kawartha Lakes, Muskoka, Parry Sound, Haliburton, Peterborough, Northumberland, Prince Edward, Hastings, Lennox and Addington, Frontenac, Renfrew, Lanark, Leeds and Grenville, Ottawa-Carleton Region, Prescott and Russell, and Stormont-Dundas-Glengarry.

Although the region includes urban centres such as Owen Sound, Huntsville, Orillia, Barrie, Lindsay, Peterborough, Kingston, Brockville, Ottawa, and Cornwall, as well as farms, roads, and other forms of development, it is mostly characterized by forests, wetlands, rock barrens, lakes, and rivers. Because central Ontario includes the southern edge of the Canadian Shield, it is often thought of as the "Gateway to the North."

The Setting for *Nature's Year*

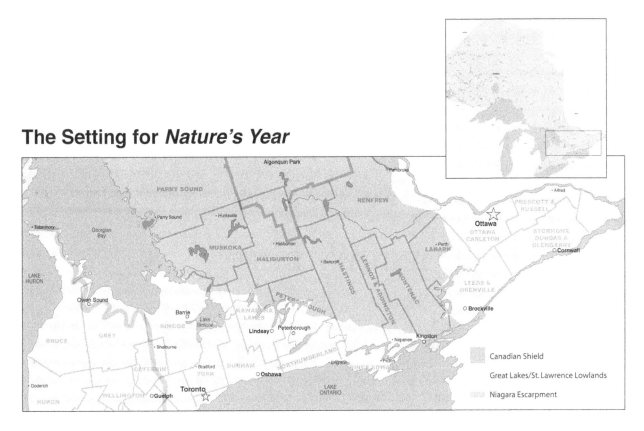

Canadian Shield

Great Lakes/St. Lawrence Lowlands

Niagara Escarpment

CLIMATE

Few places have seasons as distinct as those of central and eastern Ontario. Located in the middle latitudes — almost exactly halfway between the North Pole and the equator — the region lies between the source areas of cold (polar) and warm (tropical) air masses. These air masses regale us with weather conditions created in the Arctic or over the Gulf of Mexico. In summer, these might be the hot, muggy conditions developed in the air mass while it was sitting over the West Indies; in winter, bitter Arctic cold is more typical. In a nutshell, our climate is characterized by the interaction of these two air masses, punctuated by the steady passage of high- and low-pressure systems and a generally west to east movement of air. This leads to a great deal of variability, often from one day to the next. It also means that marked seasonal fluctuations in temperature typify the region. Compared to many other locations in the world on the same latitude, spring tends to be relatively late in central and eastern Ontario as Arctic airstreams are slow to surrender their influence. Summers are warm and humid, while winters are generally cold with significant snowfalls. Snow accumulation is greatest in the snowbelt areas near Georgian Bay. Severe summer storms are also commonplace.

Region of eastern and central Ontario covered in *Nature's Year*.
Jeff Macklin

17

As for differences within the region, nighttime temperatures usually decline from south to north and from low to high elevation. The climate in areas adjacent to the Georgian Bay area is moderated somewhat by the presence of this huge body of water. In summer, for example, the daily range of temperature — difference between the maximum and minimum — is smaller than in the rest of central and eastern Ontario. The modifying effects of the water result in a higher number of frost-free days than in most other areas. The area also experiences a higher amount of precipitation — 60 percent of which falls from November to April — and receives the most snowfall in central and eastern Ontario, with an average of 305 centimetres. On the other hand, the Renfrew County area, in the Ottawa Valley, has one of the driest climates in the region.

Land and Wildlife

Central and eastern Ontario embraces two of Canada's principal physiographic regions. It includes the southern part of the Canadian Shield, a vast area composed of ancient Precambrian bedrock, as well as the northern portion of the Great Lakes–St. Lawrence Lowland, a region of younger sedimentary rock from the Paleozoic era. As you drive north up onto the Canadian Shield, the first thing you notice is the change in the rock. Suddenly the limestone disappears, replaced with beautiful pink granite and other igneous and metamorphic rocks that border roadsides and lakeshores. With so much extremely hard, granitic rock near the surface, the soil is thin and poor. Abandoned farms attest to the difficulty of carrying out agriculture. Because farming in many areas is impractical, much of the land is forested. Satellite images clearly show a largely unbroken expanse of dark green tree cover on the Shield, interspersed with wetlands and rock barrens. However, the lowlands to the south of the Shield show only scattered pockets of tree cover. Although the northern parts of central and eastern Ontario were heavily logged in the past, they have now largely reverted to forest.

The land south of the Shield is lower in elevation and has more fertile, calcareous, loamy soils. Limestone, laid down 490 million years ago during the Ordovician period, overlies the basal Shield rock. This limestone is most visible along the 725-kilometre Niagara Escarpment, which traverses the Bruce Peninsula. The escarpment, once the bed of a tropical sea, was formed when sediments were compressed into rock, mainly limestone (dolostone) and shale. The rock was then carved by the action of glaciers and the elements into dramatic landforms such as spectacular cliffs, caves, and deep valleys. Limestone is also easily seen in road cuts all along the edge of the Canadian Shield in villages such as Marmora and Buckhorn, and near Kingston.

The Frontenac Arch is another prominent and interesting landform of central and eastern Ontario. The arch is a ridge of ancient granite rock that extends southeast from Algonquin Park, across the St. Lawrence River east of Kingston, and into New York state. It includes the Thousand Islands. The Frontenac Arch links the Canadian Shield in central Ontario to the Adirondack Mountains of New York and has always served as a vital migration route for plants and animals between these two areas. It has the greatest diversity of living things in eastern Canada.[3]

Another main feature of central and eastern Ontario is its lakes and waterways. They include the Trent-Severn Waterway, the Rideau Canal, the Rideau Lakes, Muskoka Lakes, the Kawartha Lakes, Land O' Lakes, and the lakes of the Bancroft-Haliburton area. The mix of Canadian Shield country, lowlands, and watercourses gives this area of Ontario one of the richest assortments of habitats in the province. The tree cover is characterized as Great Lakes–St. Lawrence Forest Region. Conifers such as eastern white pine, red pine, eastern hemlock, and white cedar grow in the company of deciduous broad-leaved species such as sugar maple, red maple, American beech, yellow birch, American basswood, red oak, and hop-hornbeam. However, species more common in the Boreal Forest Region to the north also exist here in good numbers. They include white and black spruce, balsam poplar, jack pine, trembling aspen, and white birch. Other common habitat types include abandoned farmland, agricultural land, and a variety of wetlands such as bogs, fens, marshes, and swamps, the latter dominated by black ash, speckled alder, or white cedar. Bogs are acidic wetlands that are low in nutrients and covered by spongy peat deposits and a thick carpet of sphagnum moss. Fens, too, are peat-forming wetlands. However, they are less acidic, have higher nutrient levels, and are dominated by sedges. Fens support a much more diverse plant and animal community.

There are also more localized habitats: the extensive bare rock, oak, and juniper ridges of places like Kawartha Highlands Provincial Park; the largely open, thin-soiled expanses of flat, surface limestone known as alvars; the pockets of Atlantic coastal plain vegetation that expanded here from the eastern U.S. coastal region during the melting of the last ice sheet 10,000 years ago; and even a few tiny remnants of tallgrass prairie, to name a few.

With such a variety of habitats, the diversity of plants and animals is one of the most extensive in the province. Some species, like Labrador tea, moose, and Common Raven are typical of northern Ontario; others, like bitternut hickory, opossum, and Cerulean Warbler are more characteristic of southern Ontario.

POTENTIAL IMPACT OF CLIMATE CHANGE

Anticipating and observing the passage of the seasons used to be a more predictable affair. The same natural events would happen more or less at the same time from one year to the next. As a result of climate change, however, many of these events — especially those happening in early spring and late fall — are becoming much more variable in their occurrence. In 2007, for example, winter did not arrive until almost February. In 2010, trilliums were blooming in the Kawarthas in mid-April, several weeks ahead of schedule.

We have always expected the weather in central and eastern Ontario to surprise us. Changeability is probably the most important characteristic of our region's climate, given our location halfway between the equator and the North Pole. However, for more and more of us, the erratic nature of the weather that is now the norm — record high temperatures in most months, severe windstorms, and either extreme precipitation or near-drought — can no longer be explained by normal variability. Although somewhat tentatively, we are beginning to point the finger at climate change to make sense of the change in weather.

How will a changing climate affect central and eastern Ontario in the long term? According to research published by the Union of Concerned Scientists about climate change in the Great Lakes region of North America, southern Ontario's climate will grow considerably warmer and probably drier during this century, especially in the summer.[4] Average temperatures are expected to rise this century from 4° to 8°C in summer and 3° to 7°C in winter. Extreme summer heat, too, will be more common, as will the frequency of heavy rainstorms. Although little change in annual average precipitation is expected, seasonal shifts are likely to occur. The overall increase in temperature may mean that Ontario will see drier soils and more droughts. Seasonally, winter precipitation is expected to increase by 10 to 30 percent, while summer precipitation is expected to remain the same. The growing season in southern Ontario could be four to seven weeks longer, as well. The ongoing decline in ice cover on the Great Lakes and inland lakes is also expected to continue.

Forests, in particular, will be impacted as the climate warms. There will probably be a reduction in the overall health of the forests because of higher concentrations of ground-level ozone and more frequent droughts and forest fires. Insect damage is projected to get worse, too, as a result of the hotter and drier summer conditions and the milder winters. A shift in tree species composition is likely to occur, with some species potentially extending their range northward over time. There is also a high probability that diversity in bird species will be affected. Resident birds like Northern Cardinals, White-breasted Nuthatches, and Black-capped Chickadees will be able to breed earlier and raise more young. However, increased resident bird populations could reduce the food and other resources available to migratory species.

As for mammals, nuisance species such as raccoons and skunks will probably benefit from milder winters. The already prolific white-tailed deer may also fare well, since this species is not really adapted to traditional Canadian winters to begin with. Moose, on the other hand, could be negatively affected, not only by the warmth but by increased levels of deer-carried parasites such as brainworm.

As lake and stream habitats warm up, native fish communities could change fundamentally, too. Cold-water species such as Lake Trout and Brook Trout will have a much harder time surviving. A changing climate may favour non-native invasive species with generalized habitat and feeding requirements over native species with more specialized needs. This will likely compound the impact of climate change in aquatic ecosystems. Zebra mussels and Common Carp, for example, may expand their range northward and, along with the introduction and spread of other invasive species, fundamentally change native fish communities. Ontario's wetlands and the species that depend upon them will almost certainly be impacted negatively by earlier spring runoff, more intense flooding, and lower summer water levels. This, in turn, is likely to reduce suitable habitat for amphibians, migratory shorebirds, and some waterfowl species.

These changes will have an important impact on how it "feels" to live in central and eastern Ontario, especially for those old enough to remember how things were. Already, to some of us, November seems more like October and April like May. One can't help but wonder for how much longer we'll be able to depend on the seasonal rituals we've kept for so long. Hunters, as much as any group, are seeing this. The November deer hunt used to be characterized by cool weather and often a light covering of snow. Now, mild weather is often the norm. The hunt does not feel the same, nor is the enjoyment. The same could be said for many other rituals, from the first lake swim of the year to the first cross-country skiing outing. The old dates for these "first of the year" events can no longer be relied upon. Although the hour is late, we can't give up the fight to convince politicians — and our families, friends, and neighbours who elect them — to take aggressive action on cutting greenhouse gas emissions, if only to set an example to the rest of the world and perhaps avoid the worst-case scenarios of a changing climate.

Resident birds like the White-breasted Nuthatch may benefit from a warming climate.
Kim Caldwell

SEASONAL OCCURRENCES OF NATURAL EVENTS OF SPECIAL INTEREST[5]

The following chart provides a summary of when some of the most noteworthy events in nature (e.g., birdsong) or nature-watching opportunities (e.g., moose-viewing) take place. Please note that there may be exceptions, depending on the geographic location in Ontario and abnormal weather patterns. The first letter is for early month, the second for mid-month, and the third for late month.

Event	Jan	Feb	Mar	Apr	May	Jun	July	Aug	Sep	Oct	Nov	Dec
Bird Feeder Activity	MMM	MMM	MHH	HHH	HMM	LLL	LLL	LLL	LLM	MHM	MMM	MMM
Songbird Migration	NNN	NNN	NNL	LMM	HHH	LNN	NNN	NLM	HHM	MLL	LNN	NNN
Waterfowl Migration	NNN	NLL	MMH	HHM	MLN	NNN	NNN	NNN	NNL	MMH	HHH	MLN
Birdsong	NNN	NNL	LMM	HHH	HHH	HHH	MLL	LNN	NNN	NNN	NNN	NNN
Amphibian Chorus	NNN	NNN	NNL	MHH	HHH	MMM	LLL	NNN	NNN	NNN	NNN	NNN
Insect Chorus	NNN	NNN	NNN	NNN	NNN	LLL	LMH	HMM	HMM	MLL	NNN	NNN
Butterfly Diversity	NNN	NNN	NNN	NLL	MMH	HHH	HHM	MML	LLL	LLN	NNN	NNN
Biting Insects	NNN	NNN	NNN	NNL	LMH	HHH	MMM	LLN	NNN	NNN	NNN	NNN
Spring Ephemeral Wildflowers	NNN	NNN	NNN	NLM	MHH	MLN	NNN	NNN	NNN	NNN	NNN	NNN
Orchids Blooming	NNN	NNN	NNN	NNN	NNL	MHM	LLL	LLM	MLN	NNN	NNN	NNN
Asters and Goldenrods Blooming	NNN	NNN	NNN	NNN	NNN	NNN	LLL	MHH	HHM	MLL	NNN	NNN
Fall Leaf Colour	NNN	NNN	NNN	NNN	NNN	NNN	NNN	NNN	LMH	HML	NNN	NNN
Moose Viewing	LLL	LLL	MMH	HHH	HHH	HHH	HHH	MMM	MMM	MMM	MMM	LLL
Beaver Viewing	NNN	NNN	NNN	NLM	HHH	HHH	HHH	HHH	HHH	HHH	NML	NNN

Level of Activity: N — none or irregular occurrence, L — low, M — medium, H — high

JANUARY

Silence and Survival

An American Robin feeding in mountain ash.
Kim Caldwell

Whose woods these are I think I know.
His house is in the village though;
He will not see me stopping here
To watch his woods fill up with snow.

— Robert Frost, *Stopping by Woods on a Snowy Evening*

For those of the natural world, January is a deadly serious time; survival is the only consideration. For many animals this means a day-to-day struggle to eat enough to simply get through the long winter night. In January, sound is the exception and silence the rule. Granted, the quiet may be broken by the Styrofoam squeal of frigid snow underfoot, by the rifle shot of swollen tree fibres bursting in the cold, or by the tinkling calls of a flock of finches passing overhead, but these sounds are simply pauses in a world of silence. Even the January moon shines with a cold-hearted light that only accentuates the stillness of the land.

But only to the casual, hurried observer is the landscape lifeless. Even on a cold, sunny morning, as frost crystals dance and dazzle in the air, a flock of chickadees may suddenly appear at the forest edge, tirelessly peering and probing for dormant insects. Nearby, a White-breasted Nuthatch works its way down a tree trunk while a Downy Woodpecker taps softly at the rough bark. In the distance, white-tailed deer browse quietly on basswood saplings, their grey winter hair matching the dim, grey-washed hues of the leafless hardwoods. The deer stop momentarily, startled by the hammer-like blows of a Pileated Woodpecker excavating a resonant old maple for dormant ants. To the curious and attentive observer, there is wonder in the countless strategies used by plants and animals to withstand or retreat from the snow, wind, and cold. Seen or unseen, awake or sleeping, life is all around us.

Birds

January Highlights

All Month Long

- Birds use a fascinating variety of strategies to survive the cold of winter. (See page 25)

- The numbers of some winter birds vary greatly from one year to the next. These species are known as winter irruptives and include many of the finches and owls. The winter of 2009, for example, saw a huge flight of Pine Siskin. (See page 26)

- In our forests, mixed flocks of foraging chickadees, nuthatches, and woodpeckers bring life to the seemingly empty winter landscape. These birds are very receptive to pishing, and the chickadees, in particular, can easily be coaxed to come in quite close. (See page 28)

- Even during the winter, woodpeckers defend feeding territories through a combination of drumming and calling.

- Pileated Woodpeckers sometimes take up temporary residence in urban areas during the winter months, providing residents with great entertainment — and some consternation — as they hammer away at neighbourhood trees in search of carpenter ants.

- Watch for Ruffed Grouse at dawn and dusk along quiet, tree-lined country roads. They often appear in silhouette as they feed on the buds of trees such as the trembling aspen.

- Large flocks of Wild Turkeys are an increasingly common sight in many areas of central and eastern Ontario. The birds are often seen in cornfields. Turkeys were reintroduced to Ontario in the 1980s.

- Bald Eagles can be seen in many parts of central and eastern Ontario in winter. They are most often found along stretches of open water (e.g., Otonabee River, Bruce Power nuclear energy plant on Lake Huron) and in areas with large winter deer concentrations (e.g., Petroglyphs Provincial Park), where they feed on the carcasses of animals that die during the season.

- Small numbers of diving ducks such as Common Goldeneye and Common Merganser can be found most anywhere there is open water, such as below dams. Large flocks of diving ducks also winter on the Ottawa River and on the Great Lakes. Watch, too, for uncommon gulls such as the Iceland and Glaucous.

- Small flocks of American Robins overwinter in many parts of central and eastern Ontario each year. When the wild fruits — grape, mountain ash, buckthorn, hawthorn, among others — that constitute the bird's winter diet are abundant, the number of winter robins increases greatly.

- Barred Owls will occasionally show up in rural backyards and prey on careless feeder birds or on mice that are attracted by night to fallen seeds under the feeder. This round-headed, hornless owl often stays for days at a time and can be quite tame.

- Large flocks of Snow Buntings can be seen all winter long in open fields and on agricultural land. Their large, white wing patches, distinctive warbling call, and unique flight behaviour immediately identify them. Snow Buntings always appear restless and continually land and take off again, swirling and veering in unison.

SURVIVING THE COLD

When it comes to winter bird survival, feathers are a bird's first line of defence. They provide efficient insulation to keep avian bodies warm, despite the cold surrounding air. A group of small muscles control each feather and can both raise and lower it. Using these muscles to fluff up their feathers, birds create tiny air spaces that drastically reduce heat loss. This is the same principle that explains why down ski jackets are so warm. Birds also have an amazing network of blood vessels in their feet and legs that minimizes heat loss. Warm arterial blood moving toward the bird's feet passes through a network of small passages in close proximity to the cold venous blood returning to the heart. The system acts like a radiator. Heat is exchanged from the warm arterial blood to the cold venous blood in such a way that heat loss is minimal.

Unless they are generating heat through flight, birds also shiver to keep warm in cold conditions. This happens even during sleep. Shivering produces heat at an amazing five times the bird's basal rate. However, when you consider that a small bird needs to maintain its core body temperature at about 42°C, and that the surrounding air may be more than 70°C colder, a great deal of fat must be burned to fuel the shivering process. Therefore, the most important line of defence for a small bird is to get enough to eat during the day in order to maintain fat reserves. A chickadee, for example, only has enough fuel to get through a single night. If it is not able to feed the following day, it will die.

Some birds can actually adjust their internal body temperature downward. This serves to reduce the difference between the bird's body temperature and the air temperature, thus further reducing heat loss. Less shivering is necessary and fat reserves are used up at a lower rate. A chickadee, for instance, can lower its core temperature from 42°C to 30°C during a long winter night. The bird actually enters a state of torpor and becomes temporarily unconscious.

The choice of the proper sleeping quarters is also important for protection from the elements. Most song birds, such as Northern Cardinals, Blue Jays, and Mourning Doves, find appropriate sleeping quarters in dense thickets of vegetation that afford shelter from the wind. Evergreens are especially popular with birds.

WINTER IRRUPTIVES

The numbers of some winter birds fluctuate widely from year to year. These species are referred to as winter "irruptives," and the years in which they are particularly common are called "flight years." Most irruptive species breed in northern Canada and winter only intermittently south of the boreal forest. Among passerines (perching birds), the main irruptive species are the Bohemian Waxwing, Cedar Waxwing, Northern Shrike, Pine Grosbeak, Evening Grosbeak, Red-breasted Nuthatch, Pine Siskin, Common Redpoll, Hoary Redpoll, Purple Finch, American Goldfinch, Red Crossbill, and White-winged Crossbill. Other irruptives include Black-backed Woodpecker, Three-toed Woodpecker, Gray Jay, Boreal Chickadee, Rough-legged Hawk, Northern Goshawk, Northern Hawk Owl, Snowy Owl, Boreal Owl, and Great Gray Owl.

The cause of this phenomenon is thought to be a shortage of food in the breeding range. This shortage follows the end of "masting." Masting refers to a year in which seed production on trees is extraordinarily high. It tends to occur over a large area so that nearly all of the trees of a given species such as white spruce or white pine are masting at the same time. The abundance of food allows birds to lay more eggs than usual and fledge more young successfully. However, in a low food year following masting, the larger than usual numbers of seed-eating birds must migrate elsewhere to avoid starving. The occurrence of masting is unpredictable and poorly understood, however. (See October plants)

Small mammals such as voles, mice, and lemmings also have cyclic population fluctuations, themselves related to the masting years. When their numbers are high, owls fledge more young than usual, but when mammal numbers crash, they must move elsewhere to find food. Snowy Owls, for example, move south when lemming populations crash on the tundra. The winter of 2011–12 was a major flight year for Snowy Owls.

In practice, the occurrence of winter irruptives is never as regular and predictable as birders would like. Winter finch invasions are especially complex. Scientists do agree, however, that the severity of the winter is not an important factor, nor do all finches respond in the same way to the same conditions. The same unpredictability is characteristic of raptor invasions. The interactions between raptors and their prey are especially complex and mammal cycles are not fully understood. It is a myth, too, that northern owls are forced south because of unusually cold weather or deep snow cover. These birds are fully adapted to winter conditions and, if prey populations are high, the owls stay on in their northern territories. The good news is that every winter sees at least several irruptive species in central and eastern Ontario.

The Great Gray Owl is probably the most impressive and visible raptor to make periodic flights into central and eastern Ontario. The largest recent invasion occurred in the winter of 2004–05 when large numbers of birds, possibly in the thousands, invaded central and southern Ontario.[1] Most of the owls were concentrated in marginal farmland interspersed with forest along the southern edge of the Canadian Shield. Their preferred prey item is the meadow vole, a species whose population crashes periodically in the boreal forest where the great gray is usually a year-round resident. This results in a mass exodus as the owls search for voles elsewhere. Unfortunately, a large percentage of these birds become traffic casualties during flight years.

The Common Redpoll is an irruptive species that often visits feeders.

Karl Egressy

PISHING: THE BIRDER'S SECRET TOOL

It is very useful to know how to entice birds in for closer observation and identification. The easiest and most effective technique is to imitate the distress call of a bird or small mammal. This technique, known as "pishing," can draw birds in like a magnet since they are curious about unusual sounds. It is also an effective way to find out if there are actually birds in a stand of conifers or similar treed habitat that might at first seem completely void of birdlife. Pishing simply involves taking a deep breath and softly but quickly repeating the word *pish* as you let the air out in one, drawn-out exhale. The noise comes mostly from the lips. The effectiveness of this technique varies somewhat depending on the species of bird but definitely works extremely well with chickadees and Red-breasted Nuthatches. These two species are almost always the first to approach in response to pishing. Be sure to continue for at least another minute, however, because other species such as Downy Woodpeckers, Dark-eyed Juncos, and Golden-crowned Kinglets are sometimes just a little slower to approach.

Ontario's Ten Most Common Feeder Birds[2]

1. Black-capped Chickadee
2. Downy Woodpecker
3. American Goldfinch
4. Blue Jay
5. Mourning Dove
6. Dark-eyed Junco
7. White-breasted Nuthatch
8. Northern Cardinal
9. Hairy Woodpecker
10. European Starling

MAMMALS

JANUARY HIGHLIGHTS

All Month Long

- Mice, shrews, voles, and moles remain active all winter long, as they make a living in the subnivean space between the ground and snow. (See page 30)

- On mild winter days when the temperature climbs above 4°C, bats sometimes emerge from hibernation and take "cleansing flights" to get rid of bodily wastes and to drink. However, bats may also appear in mid-winter because they are infected with white-nose syndrome, a disease which has now killed over a million bats in the northeastern United States, making the future for certain species very uncertain. (See page 30)

- Cougars appear to be turning up with increased frequency in Ontario. Nearly 1,000 sightings have been reported since 2002, many of them in early winter. The origin of these cats is still unclear. (See page 31)

- Moose and white-tailed deer usually shed their antlers sometime between early January and early March. This shows that the antler's role as a tool of defence is minimal, since predators such as wolves pose the greatest threat during the winter months — well after the antlers have fallen.

- Deer "yard up" in cedar swamps and large stands of hemlock such as in the Peterborough Crown Game Reserve and the west side of the Bruce Peninsula.

- Coyotes are quite vocal in the evening and at night during their January to March mating season.

- Beavers, too, mate this month or next. To tell whether a winter lodge is active, look for a dark patch, depression, or gap at the top of the lodge where the snow has been melted away by the warmth of the occupants inside.

- Chipmunks awaken regularly during the winter to make trips to their underground storehouses for food.

- Black bears give birth to hairless, sightless, and toothless cubs, no larger than chipmunks. However, they flourish on the sow's rich milk. The amount of food available in the late summer and fall is critical in determining the number of young in the litter, or even if the sow will give birth at all.

- Porcupines often take up residence in a large conifer and will spend the winter dining on the inner bark of the branches. Because of their short legs, they must avoid expending too much energy travelling through deep snow from one tree to another. Hemlock boughs scattered on the ground are often a sign of porcupine activity.

Late January
- With mating season starting, red foxes become more active. It's quite common to see their meandering trails, sometimes even in suburban backyards.

HAPPENINGS UNDER THE SNOW

Much of the mammal activity in winter is actually happening in the subnivean space under the snow. The temperature here remains just below 0°C, even when air temperatures at the surface are much colder. In this dark, damp habitat, the snow becomes crystalline and can be easily excavated by voles, mice, and shrews, with the animals forming large networks of trails. You can often see these trails in field grass and on the ground around your bird feeder after the snow melts. Little do these mammals know, however, that owls can actually hear them moving under the snow. Small mammals are especially vulnerable when in the vicinity of "ventilator shafts." These are special shafts constructed by voles in order to allow fresh air into the subnivean space when carbon dioxide levels become too high. It is amazing to watch an owl such as a Great Gray pounce on a seemingly lifeless expanse of snow and fly off with a vole in its talons. Although owls do represent a very real danger, the biggest threat to small mammals comes from the possibility of inadequate snow cover. If the snow does not arrive soon enough after the onset of subfreezing temperatures, many will perish. It is quite normal in some parts of central and eastern Ontario for snow to disappear temporarily from open areas at some point during the winter.

BAT DECLINE

As hard as it sounds to believe, there is every possibility that some of our most common bat species could completely disappear from northeastern North America in the near future. These are the bats that you might typically encounter at the cottage or in an old house or church. This dire projection is being made by Dr. Brock Fenton of the University of Western Ontario, one of the top bat researchers in the world.[3]

The freefall in bat numbers appears to be linked to a fungus that is killing the animals as they attempt to overwinter in caves and abandoned mines. Known as white-nose syndrome (WNS), biologists are now getting a clearer idea of the epidemic. The species most at risk of developing WNS is the little brown bat, although a number of other species are also vulnerable.

The most obvious symptom of WNS is the presence of a white fungus growing most often — but not exclusively — around the nose area, where it causes skin lesions. The fungus is a never-before-seen species known as *Geomyces destructans*. It now appears clear that the fungus kills the bats. In research published in the journal *Nature* in October of 2011, healthy little brown bats that were exposed to pure cultures of *G. destructans* developed WNS.[4] The results provide the first direct evidence that *G. destructans* is the causal agent of WNS. The researchers also confirmed that WNS can be transmitted from infected bats to healthy bats through direct contact. The disease is fatal to about 95 percent of the bats affected and has already killed almost six million of the animals, especially in the northeastern United States, where bat populations have been affected the longest. In 2010, WNS was detected at a number of hibernating sites across central and eastern Ontario, including a cave in the Bancroft area. It has also been detected in northeastern Ontario.

How the bats are actually dying is also becoming clearer. First of all, it is normal for hibernating bats to awaken from their torpor a few times each winter and to even fly off in search of water, especially during mild weather. However, bats with white-nose syndrome have been observed waking up far too often and staying awake too long. Being overly active during winter depletes their stored fat reserves prematurely, since the animals require a lot of energy to rouse themselves from their torpid state. To further complicate matters, there are no insects for bats to eat in winter. So, bats with WNS exhaust their fat reserves and end up starving to death.

To help curb the spread of the syndrome and minimize deaths, stay out of non-commercial caves and abandoned mines where bats may be present. If you see bats flying during the daytime in winter, or you see dead bats, please contact the Canadian Cooperative Wildlife Health Centre (1-866-673-4781) or the Natural Resources Information Centre (1-800-667-1940). People need to appreciate bats as a fascinating part of our planet's biological diversity and a stunning achievement of evolution. Their value lies in the simple fact that they exist and are here among us — at least for now.

A hibernating little brown bat with white-nose syndrome.
U.S. Fish and Wildlife Service, Wikimedia

COUGARS IN CENTRAL AND EASTERN ONTARIO

As amazing as it may sound, close to 1,000 cougar sightings have been reported to the Ontario Ministry of Natural Resources and the Ontario Puma Foundation since 2002. Many of the sightings have come from the counties of Peterborough, Haliburton, and Kawartha Lakes. Only a handful of these have been confirmed by photos, track marks, or DNA taken from scat (droppings) or other body parts. In fact, there has not been a cougar killed or captured in Ontario since 1884. Dr. Rick Rosatte, a senior research scientist with the Ontario Ministry of Natural Resources and Trent University in Peterborough, is attempting to substantiate their presence here with photographic evidence from trail cameras and physical evidence such as hair and scat samples.[5] This will help determine if the animals are wild North American cougars, or rather South American cougars that were once pets. Since the white-tailed deer is the cougar's main source of food, Rosatte believes that cougars are most likely to be in the vicinity of winter deer yards, something the sightings maps may eventually bear out. With at least 500,000 deer in Ontario, there is certainly a sufficiently large prey base for cougars to survive in this province. The cougar (*Puma concolor*), also known as the puma or mountain lion, is native to both North and South America. The eastern North American population, however, is thought to have been completely wiped out by the 1940s, mostly as a result of hunting. Nevertheless, field evidence — scat, tracks, videos, sightings, carcasses — over the past three decades suggests that cougars are once again present in their eastern North American range. They could be animals that have escaped from captivity, or animals that have been raised in captivity and intentionally released into the wild. They could also be animals that are remnants of a native Ontario cougar population. Although remote, this explanation may be possible because of the large expanses of wilderness that still exist in northern Ontario. They may also be animals that have dispersed into Ontario from other areas such as Michigan, Manitoba, Quebec, and New York State. Of course, there is the possibility that some of the sightings are of animals that are not actually cougars at all. Finally, some combination of the above could explain the sightings. Cougars, like all cats and dogs, have four toes that show. The tracks measure about eight centimetres wide. Should you see a cougar, please contact the Ontario Puma Foundation or the MNR.

FISH

JANUARY HIGHLIGHTS

All Month Long

- There is a great deal of variability in fish behaviour in the winter. Some species, such as bass, can be relatively dormant, while others, such as Northern Pike and Walleye, are active and continue to feed. (See page 33)

- Carp settle down onto the mud bottom of rivers and lakes and remain partly covered by the mud all winter.

- Anglers pursue a variety of species in winter, including Walleye, Yellow Perch, Northern Pike, Whitefish, Burbot, Lake Trout, Brook Trout, Splake, and Rainbow Trout. Central and eastern Ontario offers numerous ice fishing opportunities. Fishing is often best just after ice-up and then again at the end of the season.

- This is a period of high fish mortality, especially for young fish. During the winter months there is a shortage of appropriate-sized food such as plankton.

WINTER FISH BEHAVIOUR

Although several fish species are essentially dormant during the winter, others remain active and continue to feed. These include Black Crappie, Walleye, Yellow Perch, Northern Pike, Whitefish, Brook Trout, Lake Trout, Rainbow Trout, and Splake. As a general rule, fish in winter stay close to the bottom. Being cold-blooded, they congregate where the water is warmest and where activity requires the least energy. Many species also frequent the shallower sections of lakes. For example, Pike and Walleye are usually in less than ten metres of water. Both of these species patrol weed beds in the winter, where they hunt for perch and other forage fish. They feed most heavily at dusk, during the night, and again at dawn. Lake Trout, too, are usually found near the bottom. However, they will at times feed close to the icy ceiling of the lake. Phytoplankton levels are high here because sunlight penetration through the ice permits photosynthesis to take place. Zooplankton feed on the phytoplankton, and baitfish such as Lake Herring feed on the zooplankton. Herring in turn are a favourite food of Lake Trout. Bass, on the other hand, lie dormant under logs, weeds, or rocks until the light and warmth of spring restore their energy and appetite. Smallmouth Bass virtually starve themselves over the winter. This is one reason why so few bass are ever caught by anglers at this time of year. Bass may also aggregate in deeper areas and remain inactive.

INSECTS AND OTHER INVERTEBRATES

JANUARY HIGHLIGHTS

All Month Long
- With a little searching, it is possible to find overwintering insects in all four stages of the life cycle — egg, larva, pupa, and adult. (See page 34)

- Honey bees are one of the few insects that are able to maintain an elevated body temperature all winter. They accomplish this by clustering together in a thick ball within the hive, vibrating their wings to provide heat and eating stored honey to provide the necessary energy.

- The monarch butterflies that migrated south in late August and September are now over-wintering on 12 isolated mountaintops of the Sierra Madres, west of Mexico City in the state of Michoacán.

- Under the frozen surface of ponds, countless immature insects — larvae and nymphs — remain active. These include fierce, carnivorous dragonfly nymphs whose lower "lip" shoots out to snag prey almost like a frog tongue snatching flies, and the ingenious larvae of the caddisfly, which use bits of plants and gravel to construct protective cases around their bodies.

Honey bees have furry bodies that help to insulate them from the cold during winter.
Richard Bartz, Wikimedia

DISCOVERING INVERTEBRATES IN WINTER

A winter outing can be made all the more enjoyable by keeping an eye open for evidence of insect activity or even insects themselves. In field habitat, watch for ball-like enlargements on the stems of goldenrods, which contain the larva of the goldenrod gall fly. If you cut the gall in half with a knife, you will find the small, yellow-white larvae inside. With a little searching, you might also find the foamy egg cases of the European mantis on twigs and stems. On cherry trees, which are often found along field edges, watch for shiny brown rings about two centimetres long that encircle the twigs. They have a varnished appearance and are only slightly larger in circumference than the twig itself. These rings contain the eggs of the eastern tent caterpillar, which makes the conspicuous silken tents on cherry and apple trees in the spring. To identify cherry trees in winter, look for small trees with large black growths on the branches. The growths result from a fungal infection known as black knot. This same habitat may even reveal one of the few insects that overwinters as a pupa. The Cecropia moth spins a dense, brown, silken cocoon about five centimetres long to protect itself while it is in the pupal stage of development. The cocoon is attached lengthwise to the twigs or branches of the host plant, often a cherry, maple, or birch. In woodlands, check under the loose bark of dead trees for hibernating mourning cloak butterflies or pregnant queen wasps, both of which survive the winter by entering a state of suspended development known as diapause. (See November Invertebrates)

PLANTS AND FUNGI

JANUARY HIGHLIGHTS

All Month Long

- Winter trees and shrubs present a surprisingly wide and attractive spectrum of colours. Conifers, for example, are a study in the various greens. (See page 36)

- Twigs and buds merit special attention at this time of year. Because their characteristics are different for each species of tree, buds are a very useful tool in winter tree identification. (See page 36)

- Evergreen woodland plants such as wintergreen, pipsissewa, wood fern, and Christmas fern stand out where the snow has melted or been blown away.

- Because of their thick bunches of needles, conifer branches intercept and hold the falling snow. This results in far less snow reaching the ground underneath the trees.

- White spruce cones retain large quantities of ripe seed over the winter. This makes the white spruce a favourite food source of winter finches such as crossbills.

- The cones of red and white pine drop to the ground all winter long. The seeds in the cones, however, were already released in the fall.

THE COLOUR OF THE WINTER FOREST

To the practised eye, the January woodland is not nearly as drab as it first might appear. In fact, all of the winter trees and shrubs show surprisingly distinctive differences in colour, be it the bark or the foliage. These are not the rich colours of fall or the soft pastels of spring but they do proclaim the time of season just the same. The following list gives an idea of the colour diversity seen even in winter:

- the bronze-green of white cedar leaves
- the yellow-green of white pine needles
- the blue-green of white spruce needles
- the dark green of balsam fir needles
- the crimson of red osier dogwood twigs
- the purple hues of distant white birch crowns
- the pale yellows of lingering beech leaves and the bark of yellow birch
- the creamy-white of the bark of the white birch
- the beige-green of trembling aspen bark
- the light grey of American beech bark
- the mid-grey of white ash and sugar maple bark
- the greyish-black of black cherry bark
- the reddish-brown bark of young white cedars

BEAUTIFUL BUDS

After the dramatic leaf colour of the fall, it's easy to think that the trees are simply barren and offer nothing of interest to grab our attention. Fortunately, for those of us who enjoy winter botanizing, this is anything but the case. A closer look quickly shows that the trees are already adorned with buds — tiny jewels formed the previous summer that harbour the promise of spring; stored within the buds, the tree's entire future lies in waiting. Miniaturized, folded, and pressed together like the tiniest and tightest of parachutes, next spring's leaves are biding their time and waiting for their turn to capture sunlight and manufacture food. Embryonic stems are anxious to become twigs and eventually branches, and tiny flowers are yearning for their chance to create fertile seeds and assure a new generation of trees. But equally important, the new growth that will emerge from buds in April and May will provide food for legions of insects and small mammals that, in turn, will become fuel for the rest of the food chain. The song of a Baltimore Oriole on a May morning is directly linked to the buds of the winter forest.

Although trees can usually be identified by their overall shape and by characteristics of the bark, buds provide a more reliable means of identification, especially for the beginner. The starting point is to be able to recognize the twig, the section at the end of each branch that constitutes the previous year's growth. A twig's point of origin is marked by a distinctive ring-like node around the branch. Twigs are also smoother than the rest of the branch and almost always a different colour.

Buds can be on the side of the twig (lateral buds) or at the end (terminal buds). Because buds almost always form in the angle between the stem and the stalk of the leaf, both leaves and buds have the same arrangement on the twig. This arrangement is usually alternate or opposite. Alternate means that there is only one bud (or leaf in summer) per node. A node is a knob-like section of the twig where there is a slight fattening. A familiar example is the very apparent nodes on a bamboo cane. In opposite arrangements, there are two buds or leaves per node with one opposite the other. Honeysuckle, ash, maple, lilac, viburnum, elderberry, and dogwood are the principal tree and shrub genera with opposite leaves and buds. Just about all of the others are alternate. The following mnemonic — which unintentionally sounds like a rallying call for animal rights — may be helpful in remembering these seven genera: HAM LIVED! (Each genus, except lilac, corresponds to one letter in the mnemonic; lilac corresponds to *LI*.)

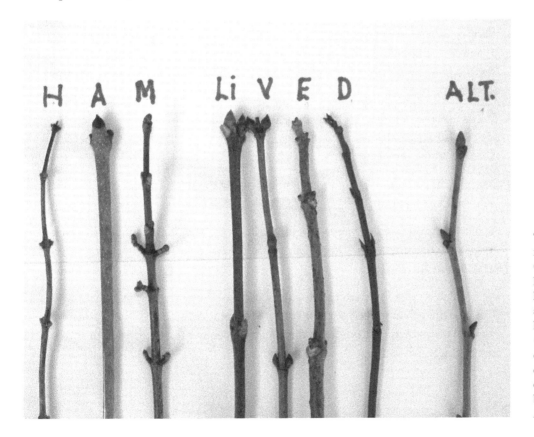

The HAM LIVED mnemonic. Opposite buds, left to right: honeysuckle, ash, maple, lilac, viburnum (highbush cranberry), elderberry, and dogwood. Alternate buds: basswood.

Drew Monkman

WEATHER

JANUARY HIGHLIGHTS

All Month Long

- Snow directly shapes the composition of our plant and animal life and is the cause of many fascinating adaptations. (See page 39)

- This is the coldest month of the year across central and eastern Ontario. Cold winter air actually originates from near or within the Arctic Circle. (See page 40)

- We begin the month with about 8 ¾ hours of daylight but gain a full hour by month's end.

- The sun actually rises later during the first week of January than it does any other time of year. In fact, it doesn't peek over the horizon until almost 8:00 a.m., depending on where you live. Compare this to June, when the sun rises at about 5:30 a.m.!

- Because there are no plants in bloom, an absence of strong smells is a characteristic feature of winter. However, our noses still know the season by the smell of everything from the smoke from a woodstove to cross-country ski wax.

- Warmer weather that sometimes arrives late in the month, usually between January 20 and 26, is known as the January Thaw. It has always held a prominent place in North American weather lore, almost akin to Indian Summer. It does not, however, occur every year.

- The daytime winter sky rarely appears as blue as the sky of summer. On cold winter days, invisible ice particles can easily form and scatter light of all wavelengths (colours), thereby making the sky appear almost white.

- Our lakes are at a uniform 4°C, except just below the ice surface, where the water is near the freezing point.

- As lake ice grows and expands, stresses are created that cause the ice to crack. The noise created can be surprisingly loud. Cracking is most frequent during periods of very cold weather.

- We become aware of the lengthening days. Most of the daylight is gained at the end of the day, rather than in the morning. Over the first three weeks of the month, we gain about 24 minutes more daylight in the afternoon, compared to only about seven extra minutes in the morning.

JANUARY WEATHER AVERAGES, 1971–2000[6]

City or Town	Daily Max. (°C)	Daily Min. (°C)	Rainfall (mm)	Snowfall (cm)	Precipitation (mm)
Owen Sound	-2.2	-9.4	23.8	110.9	134.8
Huntsville	-4.8	-15.6	18.8	83.3	102.1
Barrie	-3.2	-12.8	15.3	80.2	95.4
Haliburton	-5.4	-16.2	20	65.3	85.3
Peterborough	-3.9	-13	23.9	39.3	60.3
Kingston	-3.2	-12.2	31.5	46.1	73.5
Ottawa	-6.1	-14.8	22.9	48.7	64.2

HOW SNOW AFFECTS THE NATURAL WORLD

Snow has a huge impact on plants and animals. By forming an insulating layer over the ground, it protects certain species while making life much more challenging for others. Deep snow makes walking and feeding difficult for white-tailed deer but helps snowshoe hares to more easily reach the buds of saplings. Both plants and animals show amazing adaptations to life in a snow-covered environment. Some of these adaptations are primarily anatomical while others are behavioural. Birches, for example, have flexible trunks that bend with the weight of ice and snow without breaking. They usually spring back up when the snow melts. Snowshoe hares and weasels turn white in the winter, thereby gaining the advantage of camouflage. The hares have also developed large feet or "snowshoes" that allow them to travel on top of the snow. Ruffed Grouse grow combs on the sides of their toes which provide better support when walking on the snow. Moose, on the other hand, are equipped with long, stilt-like legs that reach down through the snow to the firm ground below while keeping the animal's belly above the surface of the snow.

Because snow and cold drastically limit food availability, many animals hibernate or sleep through much of the winter or migrate to areas where more food is available. Studies have shown that most animals can withstand the cold as long as they have access to sufficient food. Both bird and mammal migration is much more a response to the lack of suitable food than to the onset of cold weather. I have often seen large flocks of healthy American Robins flying about on frigid winter days at temperatures of -25°C or colder. As long as they can find sufficient food such as berries to fuel their bodies, they do not appear to suffer from the cold. There are also many cases in which insect-eaters that usually winter in the tropics have been able to survive an Ontario winter by dining on suet at a backyard feeder.

WINTER WEATHER'S ARCTIC ORIGINS

The next time you experience an icy blast of winter air, take a moment to reflect upon its far-off place of origin, near or within the Arctic Circle. Here, the combination of long, dark winter nights, clear skies, and a surface covered with ice and snow progressively chills the air. The sun provides little in the way of warming heat, since it only skirts the horizon for much of the winter. The snow and ice reflect away what little light the sun beams down. The thermometer eventually plummets to the frigid temperatures of the high atmosphere. As the air chills, it becomes denser and forms a huge dome of high surface pressure. As Keith Heidorn writes in his book, *And Now … the Weather*, "eventually, this cold dome breaks its bond with the spawning ground and rushes wildly southward. Howling winds precede the great air mass, announcing its advent. Trees shudder. Birds shiver."[7]

Cold Arctic air is also a necessary ingredient in snowstorms. Other than episodes of lake-effect snow (See December Weather), a typical winter storm in central or eastern Ontario is preceded by northeast winds, as a warm, moist air mass moving up from the south meets a mass of cold, dry air coming down from the north. Winter storms derive their energy from the clash of these two air masses of different temperatures and moisture levels. The point where the air masses meet is called a front. If cold air advances and pushes away the warm air, it forms a cold front. When warm air advances, it rides up over the denser, colder air mass to form a warm front. This upward movement causes the water vapour to crystallize and fall in the form of snow. In the most severe blizzards, the snow seems to fall almost horizontally, as driving winds force it through even the smallest cracks and openings.

APPROXIMATE JANUARY SUNRISE AND SUNSET TIMES (EST)[8]

(Note: Twilight starts about 30 minutes before sunrise and continues about 30 minutes after sunset.)

Location	Date	Sunrise (a.m.)	Sunset (p.m.)
Grey-Bruce	Jan. 1	8:00	4:53
	Jan. 15	7:57	5:00
	Jan. 31	7:44	5:30
Muskoka/Haliburton	Jan. 1	7:55	4:49
	Jan. 15	7:52	5:04
	Jan. 31	7:39	5:25
Kawartha Lakes	Jan. 1	7:49	4:44
	Jan. 15	7:46	4:59
	Jan. 31	7:33	5:20
Kingston/Ottawa	Jan. 1	7:42	4:37
	Jan. 15	7:39	4:52
	Jan. 31	7:26	5:13

NIGHT SKY

JANUARY HIGHLIGHTS

All Month Long

- Major constellations and stars (in italics) visible (January 15 — 8:00 p.m. EST)

  Northwest: Cassiopeia and Pleiades high in sky; Andromeda (with M31 galaxy) just above; Great Square of Pegasus setting due west; Draco just below Ursa Minor.

  Northeast: Gemini (with *Castor* and *Pollux*) high in sky; Big Dipper standing upright low in sky; Ursa Minor (with *Polaris*) to its left; Leo (with *Regulus*) rising over eastern horizon.

  Southeast: dominated by Orion (with *Betelgeuse* and *Rigel*), Taurus (with *Aldebaran*) high to its right, Canis Major (with *Sirius*) low to its left, Auriga (with *Capella*) high above.

  Southwest: an area called the "River," with large but dim constellations; Pleiades overhead.

- The winter sky provides a great opportunity to become familiar with the Big Dipper, Little Dipper, and Cassiopeia. These north sky constellations are the starting points for learning the other constellations. (See page 42)

- The Algonquian name for the full moon of January is the Wolf or Spirit Moon.

- The early winter full moon rides higher in the sky than at any other season and passes nearly overhead at midnight. Coupled with the reflective quality of snow, moonlit winter nights shine with an unforgettable brilliance. It's a great time to go for a walk or even a ski.

- Bright stars abound in the winter sky. In fact, 17 of the 33 brightest stars visible in Canada are clustered together right now in about one-tenth of the sky. They belong to a group of constellations called the "Winter Six" — namely Orion, Canis Major, Canis Minor, Gemini, Auriga, and Taurus. They also contribute to the false impression that the night sky is actually clearer in winter and that stars somehow shine brighter than in other seasons.

Early January

- The Quadrantid meteor shower takes place January 3 and 4. The point of origin is just east of the head of Draco.

GETTING TO KNOW THE NORTH SKY

It comes as a surprise to many people that the night sky changes from one season to the next. The various stars and constellations come and go in much the same way as the hummingbirds fly south in the fall and the trilliums bloom in the spring. Exactly where in the sky you will see a given constellation, however, depends not only on the time of year but also on the hour of the night. The first step in learning the main stars and constellations is to become familiar with Ursa Major (Big Bear) and its asterism, the Big Dipper. An asterism is a prominent pattern or group of stars, typically having a popular name but usually smaller than a constellation. Ursa Major, Ursa Minor (Little Dipper), and Cassiopeia (the "W") are known as circumpolar constellations because they appear to rotate around the North Star (Polaris) that is located almost exactly at true north. They will guide you to other constellations because they are above the horizon and visible all year long. Most constellations disappear below the horizon for part of the year. There is no use looking for Orion in late spring or Leo in November in central or eastern Ontario.

Ursa Major, with the Big Dipper, can always be found in the northern sector of the sky. The Dipper consists of seven stars and forms the tail and body of the Bear. To see the rest of the bear, look for two curves of stars. The curve in front of the Dipper's bowl forms the head and forepaw of the bear, while the curve below the rear of the bowl forms the hind leg. The handle of the Dipper is the bear's tail. The front two stars of the Dipper's bowl are known as the Pointers because they point directly toward Polaris, the North Star. Starting at the Pointer star at the bottom of the bowl, imagine a line connecting to the upper Pointer. Project this line five times and you will have found Polaris. There are no other bright stars near Polaris. This technique works at any time of the night and at any season of the year.

Although Polaris is not particularly bright, it is unique. When you are facing Polaris, you are facing almost due north. South is therefore directly behind you, east is to your right and west is to your left. At our latitude, Polaris is also approximately halfway up in the sky. It appears to remain stationary while Ursa Major and the other constellations seem to rotate around it. Polaris is also the first star in the Little Dipper's handle. This is helpful to know because the Little Dipper is a rather dim asterism.

The other well-known circumpolar constellation is Cassiopeia. Its five bright stars form an easy-to-recognize M or W shape. To find Cassiopeia, imagine a line that starts where the Big Dipper's handle joins the bowl, extends through Polaris, and continues the same distance to Cassiopeia. The Inuit imagined this constellation as a pattern of stairs sculpted in the snow.

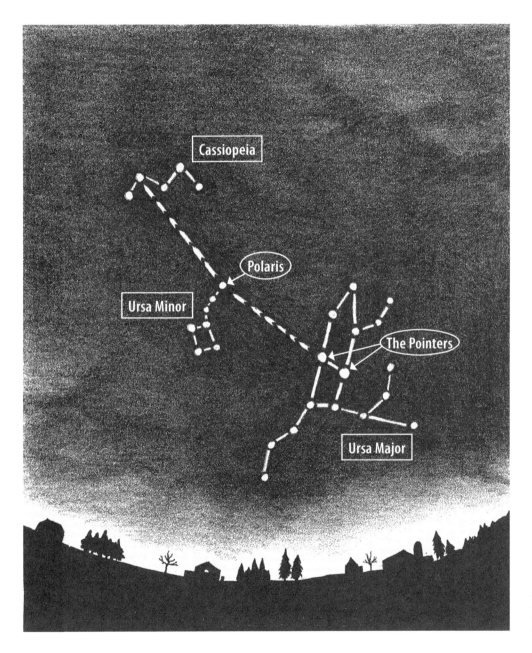

The north sky in January.
Jean-Paul Efford

FEBRUARY

Gateway to the Year

A male striped skunk on the prowl.
Kim Caldwell

Every gardener knows that under the cloak of winter lies a miracle …
a seed waiting to sprout, a bulb opening to the light, a bud straining to
unfurl. And the anticipation nurtures our dream.

— Barbara Winkler

FEBRUARY OPENS TO WHAT IS OFTEN THE COLDEST WEEK OF THE WINTER. ICE, SNOW, and cold temperatures usually reign supreme. Ironically, this is also the time we amuse ourselves with stories of a sleepy-eyed rodent emerging from its hole to gauge the prospects of spring. Although we cannot take the story literally, Groundhog Day does mark the mid-point of winter. In earlier days it was a time to take stock that you had half your hay, root crop, and firewood left in order to comfortably make it through to spring.

February is maybe the best month to enjoy winter. Sunset is not until 5:30 p.m. which allows for long afternoons to spend outdoors. Snow cover is more substantial as well. For some, there is the romance of the "big winter storm." It starts with an atmosphere of anticipation and the prospect of real excitement. There is the exhilaration of being out in the wind and snow and, afterward, a spirit of camaraderie as friends and neighbours help each other dig out.

But winter can begin to weigh heavy on our psyche this month. At first glance, spring doesn't appear to be coming any time soon. Yet no matter how well-entrenched snow and ice may seem, the promise of spring does begin to reveal itself this month. Forget what the calendar is saying and simply look, listen, and smell. Already, there is a very noticeable stretch in daylight, American Crows and Horned Larks are returning, and the first real birdsong since last summer can be heard. Starlings are piping and cackling, chickadees are whistling their familiar song, and even cardinals are singing once again, caught up in the urge to reaffirm their pair bonds. Our noses, too, are alerted to the coming change of season, often by the smell of a skunk out searching for a mate on a damp February night. And, in only a few more weeks, Red-winged Blackbirds will have returned to local marshes and the first pussy willow catkins and wild leeks will have emerged.

BIRDS

FEBRUARY HIGHLIGHTS

All Month Long

• Although rather tentative at first, birdsong returns in February as pair bonds are established or renewed. Black-capped Chickadees, Northern Cardinals, and House Finches are several of the birds that usually start singing this month. (See page 48)

• Highly nomadic Red and White-winged Crossbills commonly sing and, given sufficient food supplies, even nest in the middle of winter. Crossbills are most common in central and eastern Ontario in years with big spruce cone crops such as 2008–09.

• Woodpeckers begin to drum more frequently to advertise ownership of territory and to renew or establish pair bonds.

Early February

- Horned Larks return. They are traditionally the first migrant arrivals of the year and mark the coming of "pre-spring." Along with Snow Buntings, these are birds of open, agricultural land. Their tingling song is given in flight and can be heard even on the coldest days.

Mid-February

- This is courtship time for ravens. Males begin their aerial nuptial displays, diving and twisting like corkscrews over Canadian Shield country. Pairs of ravens will also soar together. There has been a dramatic expansion of the raven's range southward in recent years.[1]

- The Great Backyard Bird Count (GBBC) takes place over four days. The GBBC engages birdwatchers of all levels of expertise to create a real-time snapshot of the whereabouts and relative abundance of North American birds in mid-winter. Go to *www.birdcount.org* for details.

Late February

- "Giant" Canada Geese that have wintered along the Great Lakes and southward into the northern and mid-United States begin arriving back in central and eastern Ontario. Other species such as the Cackling Goose (formerly thought to be a sub-species of the Canada Goose) sometimes mix in with these flocks, too.

The male Common Goldeneye puts on a very vocal courtship display.

Karl Egressy

- Crow numbers increase as returning migrants bolster the ranks of those birds that never left. Watch for long, scattered flocks often flying at high altitudes.

- Common Goldeneye and Common Merganser numbers increase as migrants from the Great Lakes begin to head north. The male Common Goldeneye puts on quite a show of courtship behaviour in late winter. He thrusts his head forward and then moves it back toward his rump with his bill pointing straight up in the air, at which time he utters a squeaky call.

- Exceptionally warm weather may result in short-distance migrants like American Robins and Red-winged Blackbirds arriving back early from their wintering grounds in the southern United States. Waterfowl may arrive, as well.

RENEWING PAIR BONDS

For many of our resident birds, the first step in the breeding cycle begins in February with the formation or reaffirmation of a pair bond between a male and female. With long-lived species such as geese, cranes, and some owls, the same pair of birds may remain together in a monogamous relationship for life. For other species, a pair may stay together for several successive breeding seasons (e.g., Mourning Doves), for one entire breeding season (most songbirds), or for only a single nesting (e.g., House Wrens). In some species, such as American Robins, pairing arrangements seem to vary with individual birds.

Even in the case of birds that have a lifelong pair bond, the attachment between the male and female weakens during the fall and early winter. The birds may be indifferent or even aggressive to each other during this time. By late winter, however, a reaffirmation of the pair bond begins. This is especially apparent in cardinals. Although the male and female remain together throughout the year, the bond is relaxed in the non-breeding season. But starting in February, the male starts to sing regularly once again and both sexes often feed together. The male White-breasted Nuthatch is also becoming much more interested in the female with which he shares the same territory throughout the year. He sings in her presence, bows before her while fluttering his wings and cocking his tail, and may even bring her food. Male Black-capped Chickadees, too, whistle their three-note song much more often starting in mid-winter. This serves to attract a new mate or to strengthen the pair bond with an existing partner, since some chickadees actually remain mated for life. Chickadee flocks begin to break up as spring approaches, and pairs take up residence on nesting territories. An easy mnemonic for the chickadee's love song is "Hi Sweetie," which coincides nicely with both the date and the meaning of Valentine's Day.

SECRET LIVES OF CHICKADEES

It's hard not to love chickadees and their ability to survive the long, cold winter nights, bitter winds, and relative lack of food. However, there is also a lot going on in the chickadee world that is not immediately obvious. Chickadees live in a surprisingly complex social system, based on a dominance hierarchy, or "pecking order."[2] Each bird in the flock is known to the others according to its rank. The rank, in turn, depends on the bird's degree of aggressiveness. All of the birds in the flock are subordinate to the most aggressive bird. Males dominate females and younger birds are usually subordinate to older ones. Dominance can be expressed through singing, body position, chasing, and sometimes even by physically fighting — as a friend of mine once witnessed. High rank in the dominance hierarchy confers some very important advantages. For starters, dominant birds enjoy the best access to food. At the feeder, for example, the dominant male can easily frighten other chickadees away.

If you are the number one bird in the flock, you can also count on your mate always being faithful. A lower-ranked male, however, is often cuckolded by his partner, who tends to look for sexual opportunities with a higher ranked bird. The female who is paired to the alpha male also enjoys better access to food and very little aggression from other birds. She, too, has a higher overwinter survival rate. When the nesting season arrives, she will also lay more eggs than lower ranked females and suffer less nest predation. When chickadees come to your feeder, watch for short chases between members of the flock. This is an expression of dominance that is most often seen when the feeder is crowded. Dominant birds will approach the feeder directly, scaring off others that are feeding. Lower-ranked birds tend to wait until the more dominant individuals are through. You will often see them approach the feeder and then veer off without landing.

Black-capped Chickadees live in a surprisingly complex social system.
Karl Egressy

MAMMALS

FEBRUARY HIGHLIGHTS

All Month Long

- Watch for river otters in winter around areas of flowing water such as streams and rivers. Their trough-like "snowslide" trails are sometimes seen on embankments or even flat ground.

- Being very social animals, northern flying squirrels sometimes join up in single-sex groups for warmth during the winter. They will often choose a tree cavity or abandoned woodpecker hole. Watch for them at night at bird feeders, too.

- Gray squirrels mate in January or February and can often be seen streaming by in treetop chases as a group of males pursue a half-terrorized female. Some amazing acrobatics are usually part of the show. The female gives off a scent at this time of year that the opposite sex finds irresistible.

- The Virginia opossum, a marsupial that is native to the southeastern United States, is extending its range northward into parts of central and eastern Ontario, probably as a result of factors such as climate change and the resulting milder winters. These mostly nocturnal animals are sometimes attracted to bird feeders in winter, where they eat spilled seed.

- Bats that are overwintering in attics and other dry building environments will sometimes emerge from hibernation to try to find water. You may also see them during periods of extreme cold, which can awaken bats hibernating in attics. They may try to squeeze through the vapour barrier in the wall to reach warmer temperatures and will sometimes crawl down between walls and find their way into basements. If you immediately release them outside, however, they will freeze to death.

Early February

- Don't waste your time looking for groundhogs. No sane "woodchuck" is out of hibernation yet, and the last thought in their mid-winter dreams is whether the February sky is clear or cloudy!

Mid-February

- Male striped skunks and raccoons roll out of their dens any time from mid-February to early March and go on a long prowl looking for females with which to mate. (See page 51)

Late February

- In years of deep snow, late winter is the most difficult part of the year for white-tailed deer. (See below)

- Wolves mate between late February and early April. Only the alpha male and female in the pack will reproduce.

MAMMALS ON THE MAKE

The so-called "dead of winter" is anything but dead for many of our mammals. Many species must mate at this time for their young to be born in April or early May when food becomes more plentiful and conditions in general are more conducive to raising a family. This is the case for some of our best known mammals — gray squirrels, eastern wolves, coyotes, red foxes, raccoons, striped skunks, and American mink — all of which have a gestation period averaging about two months. Because mammals tend to be so secretive, most mating activity goes on without our knowledge. There is one indicator, however, that everyone knows — the smell of a skunk on the prowl on a mild February or March night. The smell of a skunk is a time-honoured sign of pre-spring and one of the first datable natural events of the year recorded by early phenologists — ecologists who study periodic plant and animal life-cycle events that are influenced by climate and seasonal change — such as Aldo Leopold.[3] Skunks often overwinter communally in groups of up to 12 individuals, most of which are females. Being polygamous, the male mates with all the females in the den before heading out to look for love elsewhere.

WHITE-TAILED DEER IN WINTER

Deer do not cope especially well with winter. In Ontario, white-tailed deer are at the northern fringe of their range. They lack the anatomical adaptations such as the long legs that moose have developed. When there is more than 50 centimetres of soft snow on the ground, deer are seriously handicapped and must expend large amounts of energy simply walking from one point to another. They have, however, adapted in terms of their behaviour. Every year, most deer move to special wintering areas where the snow cover is thinner and browse is available. In agricultural lands, this wintering site may be a cedar swamp or flood plain or simply the south-facing slope of a woodlot or field. In more northern locations, many deer move into larger sites known as "deer yards." These are areas of mostly coniferous tree cover which offer browse, protection from the wind, and less snow to contend with because much of it remains caught in the branches.

The winter diet of deer consists mainly of deciduous buds and twigs, conifer branches, and both coniferous and deciduous saplings. Some of the preferred species include white cedar, eastern hemlock, American basswood, and apple. If corn is available, it is also a very popular food item. In deer yards, however, most of the new twig growth on woody shrubs may have been consumed by late winter. The animals sometimes end up relying on older twigs which have a much lower food value.

Surprisingly, a deer's food demands decrease during cold weather, because its metabolic activity slows down. During extended cold periods, deer will enter a state of semi-dormancy, sometimes called "hibernation on the hoof." This is an adaptation to survive the relative lack of food and the adversity of winter weather. Food demands can decrease by 30 percent. The animals will also bed down for much longer periods of time.

Even with these adaptations, if appropriate browse becomes too scarce, pregnant does will actually resorb the developing fetus into their body, and no fawn will be born in the spring. Weakened deer may also fall victim to coyotes and wolves or even to dogs that are allowed to run loose. This is a serious problem during winters with deep, prolonged snow cover and when the snow is crusty. In years of severe food shortage, and only as a last resort, humans sometimes provide deer with supplementary or emergency feeding. Contact the Ministry of Natural Resources before going ahead with such measures. Providing the wrong type of food can actually kill the deer.

White-tailed deer show a number of behavioural adaptations to winter.
Terry Carpenter

FISH

FEBRUARY HIGHLIGHTS

All Month Long

• Throughout February and into early March, Burbot spawn at night under the ice on rocky lake bottoms. Also known as Ling or Freshwater Cod, the fish form writhing balls of about a dozen intertwined individuals which actually move across the lake bottom. Burbot are found in deep water bodies throughout central and eastern Ontario. Ice fishermen sometimes catch them after dark in the same winter habitats as Walleye and Lake Trout.

• Lake Trout eggs hatch in February but the fry remain in the substrate for about six weeks, surviving on energy stored in their yolk sac. They swim up from the shoal where they hatched in late March and early April and must actively feed or starve to death. Food is usually abundant at that time. However, if climate change causes an increase in water temperature, this would result in the eggs hatching earlier. The fry may then become active in mid-winter when there is no food available and consequently starve to death.[4]

INSECTS AND OTHER INVERTEBRATES

FEBRUARY HIGHLIGHTS

All Month Long

• On mild, sunny days, check the snow along the edge of woodland trails for snow fleas. What looks like spilled pepper might begin to jump around right before your eyes! (See page 54)

• Wingless scorpion flies and wingless winter crane flies can sometimes be found walking on the snow on cloudy winter days with mild temperatures. The latter is very slow moving, has long legs, and is amazingly spider-like in appearance.

• True spiders may also emerge from the leaf litter and head out onto the snow pack on mild winter days. These are spider species that overwinter as adults and do not build webs to catch prey. They include wolf and dwarf spiders. The latter, at only 0.6 centimetres long, is often the most abundant winter spider.

Late February

- In late February and March, winter stoneflies emerge from streams and walk about on the adjacent snow in their quest to find a mate. They are black, about one centimetre in length, and have two prominent tail-like appendages. After mating, the females return to the stream to lay their eggs. The presence of stoneflies indicates good water quality.

- Spring has sprung for overwintering monarch butterflies in the mountains of Mexico. Now that the lengthening days have triggered the final development of their reproductive systems, male monarchs are zealously courting the females. Most of the males die shortly after mating.

SNOW FLEAS: INSECTS ON ICE

Even people who don't like winter will usually admit that "at least there are no bugs." Well, guess again. Insects are indeed out and about in the winter woods. The good news, however, is that the species you're most likely to encounter — the minuscule snow flea (*Hypogastrura nivicola*) — prefers algae, bacteria, and mouldy leaves to human blood. The snow flea is not related to true fleas such as those your dog or cat may bring home. It belongs to an ancient group of wingless insects called Collembola, commonly known as springtails because of their amazing ability to jump. Less than two millimetres in length, snow fleas accomplish their incredible leaps thanks to two tiny tail-like appendages — visible through a good magnifying glass — that are folded under the abdomen and held in place by a kind of hook. When the hook is released, the tails act as a catapult, sending the insect rocketing skyward. They can, in fact, hurdle themselves an impressive 13 centimetres, which represents 65 times the insect's body length. In human terms, that's like being able to jump the length of a football field.

Late winter is the best time to look for snow fleas. On mild, sunny days, carefully examine the snow adjacent to open ground where leaves and soil have been exposed through melting. The base of a tree along the edge of a woodland trail is a good place to try. What may initially appear like pinhead-sized particles of soot or pepper scattered on the snow will suddenly start jumping about in front of you. Now, you may wonder what snow fleas are doing out on the snow in the first place. Surprisingly, they are out looking for food in the form of microscopic organic matter. Snow fleas are able to remain active because of the fact that they are black, which allows them to absorb heat from the sun. The microclimate in the sheltered spaces between the ice crystals is also substantially warmer than the surrounding air. At night, they return to the shelter of the leaf litter.

PLANTS AND FUNGI

FEBRUARY HIGHLIGHTS

All Month Long

- On mild winter days when sufficient moisture is available, lichens are able to carry out photosynthesis and to actually grow. (See below)

- The grey, brittle remains of last summer's roadside wildflowers add interest to a winter walk. Watch especially for the cup-like "bird's nest" of Queen Anne's lace; the large pods of milkweed, some of which still contain silken seeds; and the lovely, almond-shaped seedcases of evening primrose, whose ends curl back like the petals of a woody flower.

- Young American beech trees retain many of their tan, papery leaves for the entire winter. They are especially beautiful on a winter morning when the sun's rays pass directly through them.

- Our sense of smell can be useful in winter tree identification. Break a twig or crush a bud of yellow birch and a wintergreen or peppermint smell is released. An equally pleasant but much stronger smell comes from the large, terminal buds of the balsam poplar. They give off a familiar spicy fragrance that immediately evokes the smell of a damp late-May morning.

- Several species of shrubs still have fruit "on the vine," even in mid-winter. These include highbush cranberry, staghorn sumac, and maple-leaved viburnum. Because the fat content of these fruits is quite low, they are seldom eaten by birds or squirrels until all other food sources fail.

LICHENS

Of all the conspicuous organisms in the natural landscape, lichens are probably the most overlooked. However, with far fewer species around to compete against for the eye's attention, winter can be a good time to get to know this interesting division of the fungi kingdom.

Lichens are actually "dual organisms" consisting of an alga and a fungus living together as a single unit for mutual benefit. Although they both can live on their own, they seem to do better together. This type of relationship in nature is known as mutualistic. The fungus, which is the visible portion of the lichen, provides the alga with protection and a "house" to live in. It also supplies the alga with mineral nutrients and water, both of which the fungus absorbs from the surrounding substrate and directly from the air. The alga, in turn, makes food for the fungus by using sunlight to photosynthesize glucose.

Although lichens have no roots, they do have fungal strands that attach the undersurface of the lichen to the tree or rock. Lichens survive the cold of winter by drying out to the point of becoming brittle. If temperatures climb above freezing, however, and if sufficient moisture becomes available, they can photosynthesize and grow even in winter.

Some of the most common lichens in central and eastern Ontario are those belonging to the genus *Parmelia*. They are pale grey or light green leaf lichens that typically grow on trees, logs, and rocks. Some species of *Parmelia* are pollution-tolerant and can easily be found on the bark of urban trees. These lichens are used by Ruby-throated Hummingbirds and Eastern Wood-pewees in the construction of their nests. A winter ski or snowshoe outing along a rock ridge or through a conifer swamp can be a good time to look for some of the other common lichens. These include lungwort, rock tripe, and the familiar British soldiers.

Ecologically, lichens are important because they often occupy niches where nothing else will grow. Species such as rock tripe, for example, are the only organisms that will grow on bare rock. Over the course of many years, the lichen will slowly collect around and beneath itself small amounts of moisture and mineral and organic fragments. When freezing temperatures come, the lichen's collected water will expand as it forms ice. This expanding action may break off a few more particles from the rock below the lichen, thereby helping to make more soil. Eventually, other more complex plants such as mosses or ferns may take root in the modest accumulation of soil and replace the lichen. *Forest Plants of Central Ontario* is an excellent guide to the common lichens. (See Bibliography)

Parmelia caperata **is a common, light-green lichen found on the bark of deciduous trees.**
Drew Monkman

Weather

February Highlights

All Month Long

- We begin the month with about 9¾ hours of daylight and end with 11, a gain of about 75 minutes.

- Some people claim they can smell an approaching snowstorm. Snow can, in fact, contain chemicals such as formaldehyde and sulphate, all of which have distinctive odours that can sometimes be perceived by the human nose.

- In terms of total precipitation (rain and snow combined), February is one of the driest months of the year.

- February is often a particularly good time to see "sun dogs." They consist of two short rainbow arcs forming on either side of the sun and creating a halo of light. Sun dogs appear when the sun shines through a thin cloud of ice crystals, usually on days with feathery cirrus clouds.

- Central and eastern Ontario rivers such as the Otonabee sometimes flood, even in mid-winter. Cold weather conditions supercool (reduce the temperature to below 0°C) turbulent water as it rushes downstream and over dams, thereby creating frazil ice. This type of ice is characterized by loose, needle-shaped ice crystals which resemble slush. Frazil ice has the ability to "stick" to the upstream side of objects in the water and grow in size as more ice gets deposited. This can create ice jams which impede the flow of the river and cause flooding.

Early February

- Groundhog Day, February 2, marks the mid-point of winter. However, in the real world, no self-respecting groundhog will be out of hibernation until at least mid-March. (See page 58)

Late February

- Winter is on the ebb. The sun is rising and setting farther and farther north with each passing day. This means that it travels higher through the sky, making for longer days.

- Daylight is with us now until close to 6:00 p.m. The increase in daylight rekindles the human spirit!

FEBRUARY WEATHER AVERAGES, 1971–2000[5]

City or Town	Daily Max. (°C)	Daily Min. (°C)	Rainfall (mm)	Snowfall (cm)	Precipitation (mm)
Owen Sound	-1.7	-9.5	19.9	58.1	78
Huntsville	-2.7	-14.4	12.4	54.5	66.9
Barrie	-2	-12.1	13.3	39.5	52.8
Haliburton	-3.3	-15	14.4	52	66.4
Peterborough	-2.2	-11.8	21.4	30.5	49.7
Kingston	-2.6	-11.7	28.1	37	62.1
Ottawa	-6.1	-14.8	22.9	48.7	64.2

GROUNDHOG DAY

Even though the first week of February is often one of the coldest weeks of the winter, tradition steers our thoughts to spring at this time and to the escapades of the groundhog. This familiar member of the squirrel family is supposed to arise and to rule on the arrival date of spring. If it sees its shadows, the timid animal will immediately dive back into its burrow and winter will continue for six more weeks. If there is no shadow, spring will soon be at hand. Unfortunately, as most people know, the groundhog is a true hibernator and sleeps right through the whole event. The Groundhog Day tradition is linked both to Roman times and to the Christian celebration of Candlemas Day. In mid-winter, the Romans burned candles to Februa, the mother of Mars, asking for her intercession in bringing about an early, clement spring. The burning of candles later became part of the Christian Candlemas celebration, which also takes place on February 2. Although it commemorates the purification of the Virgin Mary, Candlemas has long been associated with the idea of forecasting the coming of spring. An old English poem states:

> If Candlemas Day be fair and bright,
> Winter will have another flight.
> If Candlemas Day be cloud and rain,
> Then winter will not come again.

Many early peoples believed that animals had the ability to forecast the weather. They therefore looked for portents to tell if spring would be early or late. The badger became the Candlemas forecaster in England. Although European badgers are dormant much of the winter, they do occasionally come out of their "setts" and can be seen grooming themselves. It is therefore likely that people would have seen badgers at this time of year. It is also probable that the early English settlers who came to North America mistook the groundhog for a type of badger and simply transferred the weather forecasting tradition onto the shoulders of our sleep-loving rodent.

There is no proof that any animal or plant behaviours exist that can alert us days or weeks ahead of time as to whether the coming season will be early or late, or abnormally warm or cold. However, as compared to humans, some animals are definitely able to make better use of their five senses to detect changes in the environment that accompany impending weather phenomena. For example, some birds can sense a drop in barometric pressure and therefore instinctively seek cover before a rainstorm. No long-term weather prediction ability has ever been demonstrated, however, be it a woolly bear caterpillar or a groundhog!

In late summer, the flower head of the Queen Anne's lace curls inward to a form a cup-like "bird's nest." This makes the plant easily identifiable and quite attractive all winter long. *Wikimedia Commons, Vera Buhl*

APPROXIMATE FEBRUARY SUNRISE AND SUNSET TIMES (EST)[6]

(Note: Twilight starts about 30 minutes before sunrise and continues about 30 minutes after sunset.)

Location	Date	Sunrise (a.m.)	Sunset (p.m.)
Grey-Bruce	Feb. 1	7:43	5:31
	Feb. 15	7:25	5:51
	Feb. 28	7:04	6:09
Muskoka/Haliburton	Feb. 1	7:38	5:27
	Feb. 15	7:20	5:46
	Feb. 28	6:59	6:04
Kawartha Lakes	Feb. 1	7:32	5:21
	Feb. 15	7:14	5:41
	Feb. 28	6:53	5:58
Kingston/Ottawa	Feb. 1	7:25	5:14
	Feb. 15	7:07	5:34
	Feb. 28	6:46	5:51

NIGHT SKY

FEBRUARY HIGHLIGHTS

All Month Long

- Major constellations and stars visible (February 15 — 8:00 p.m. EST)

  Northwest: Pleiades high in west; Cassiopeia in mid-sky; Great Square of Pegasus at horizon; Andromeda (with M31 galaxy) just above it.

  Northeast: Gemini (with *Castor* and *Pollux*) near zenith; Auriga (with *Capella*) to its left; Big Dipper standing upright low in the sky; Little Dipper (with *Polaris*) to its left; Leo (with *Regulus*) to its right.

  Southeast: Gemini near zenith; Canis Minor (with *Procyon*) below Gemini; Canis Major (with *Sirius*) due south.

  Southwest: dominated by Orion (with *Betelgeuse* and *Rigel*); Taurus (with *Aldebaran*) high to its right; Canis Major (with *Sirius*) low to its left; Auriga (with *Capella*) high above.

- The Algonquian name for the full moon of February is the Snow Moon.

- Leo, the constellation of spring, holds sway over the early northwestern morning sky as we head for work. Its promise of spring somehow makes the cold and darkness easier to endure.

- Because the waxing crescent moon is fairly high above the horizon at sunset in late winter and early spring, this time of year provides the best showing of "earthshine" — the faint lighting of the dark portion of the moon during the waxing crescent moon phase. It is caused by sunlight reflected off the earth, onto the moon, and back again to our eyes. Watch the western sky just after sunset, two to four days after the new moon.

A waxing crescent moon, showing "earth-shine." Note Venus to the lower right.

Rick Stankiewicz

MARCH

Waiting for Spring-to-Be

LESSER SCAUP

PIED-BILLED GREBE

WOOD DUCK

HOODED MERGANSER

Some common water birds during migration.
Kim Caldwell

March comes, a kind of interregnum, winter's sovereignty relaxing,
spring not yet in control.

— Hal Borland

MARCH IS A TIME OF WAITING — WAITING FOR SPRING-TO-BE; HOWEVER, OUR patience is starting to wear thin. Whenever we think that winter is finally ready to give up her reign, we are hit by yet another blast of cold and snow. March is essentially a tug-of-war between winter and spring. And anything and everything goes — wind, snow, frigid cold, freezing rain, thundershowers, near-summer warmth. For this seems a month without rules.

March rarely flaunts the signs of spring. Many signs are subtle and reveal themselves only to those who actively search them out. But despite what the day-to-day weather may be doing, many indications of the impending change of season simply can't be missed. The sun now is higher, brighter, and warmer; we have daylight from 6:00 a.m. until early evening; Leo, the constellation of spring, looms high in the east; the daytime sky is full of noisy flocks of crows, and local wetlands are brightened by pussy willow buds as they reveal their insulating tufts of white hairs. The first true spring migrants are only a mild spell away, since a period of warm weather will bring in a flood of ducks and at least half a dozen species of songbirds.

In March, our sense of smell is reawakened by the delicious aroma of boiling maple sap and by the pervasive odour of the thawing earth. Our ears are reacquainted with birdsong as Red-winged Blackbirds, American Robins, and a dozen other migrants and resident birds advertize ownership of nesting territories. All our senses tell us that the warmth and intense activity of "high spring" are coming. But let's not rush the season. Each stage of spring's arrival needs to be savoured, because it will be over all too soon.

BIRDS

MARCH HIGHLIGHTS

All Month Long

- Open sections of lakes and rivers are host to thousands of northward-bound ducks, impatiently awaiting the departure of the ice. (See page 65)

- March is usually the best time of the year to hear a variety of different owls. Barred Owls scream, hiss, hoot, and cackle from mixed forests on the Canadian Shield. Farther south, Great Horned Owls hoot and Eastern Screech Owls whinny from woodlots in mostly agricultural areas. Throughout central and eastern Ontario, you may also hear the single-note whistle of migratory Northern Saw-whet Owls. It is repeated monotonously without any interruption.

- The first migratory songbirds return this month and birdsong increases accordingly. If you don't already know the songs of these common birds, this is a great time to start learning them. A great website to try is *www.birdjam.com/learn.php*. (See page 64)

- Meltwater ponds often form in low-lying sections of agricultural fields. These ponds often welcome both Canada and Snow Geese as well as a variety of interesting ducks such as Northern Pintails, American Wigeon, and Wood Ducks. In the Cornwall and Ottawa areas, Snow Geese are abundant in flooded fields.

- Starlings and House Sparrows are already laying claim to nest boxes. Pairs of noisy Canada Geese are staking out nesting territories on still-frozen ponds. Common Ravens and Gray Jays are busy incubating their eggs.

- Bohemian Waxwings are often present in large numbers over much of central and eastern Ontario. Some flocks number in the hundreds.

Early March

- This is the only time of year that spotting a robin generates real excitement. There is, however, an irksome question. Is the bird a true migrant or simply an over-winterer? The situation is especially complicated in central and eastern Ontario, where small numbers of robins regularly spend the winter. However, migrant or not, an early March robin is always a welcome sight.

- Great Horned Owls are usually on their eggs by now. Biologists believe that early nesting allows the young owls to be ready to learn their hunting skills in late spring when prey is most plentiful and the majority of prey animals are young themselves and therefore inexperienced at avoiding enemies.

Mid-March

- Flocks of Red-winged Blackbirds are now returning to the still-frozen wetlands of central and eastern Ontario. These early arrivals typically perch on the highest branches of trees and are easy to see.

- Red-tailed Hawks are making their way back to northern Ontario and Canada from wintering grounds in the eastern United States. Watch also for pairs of mostly resident Red-tails soaring together over their woodlot territories.

- The first northward-bound Turkey Vultures are usually seen about now. Small flocks will sometimes roost in city conifers and fly low over surrounding neighbourhoods.

- One of the most notable avian arrivals is the Common Grackle. Grackles are foot-long, glossy purple blackbirds that make a loud *chack* call as they fly around neighbourhoods in small flocks. They have a long, wedge-shaped tail.

- Bald Eagles lay their eggs between now and early April. This is followed by a 36-day incubation period. Eagles returned to Peterborough County in 2005 as a nesting species, after an absence of more than 100 years.

- In southeastern Ontario, Snow Geese are usually common in flooded grain fields by this time. Abundant field grains (e.g., corn, soybeans) along this species' migration routes and on wintering sites has contributed to a huge population increase. Coupled with the bird's feeding behaviour, which uproots and kills vegetation, the overabundance of Snow Geese is having a detrimental effect on the fragile Arctic ecosystem where they breed. To help reduce Snow Geese numbers, a special spring hunting season is being implemented in southeastern Ontario, effective in 2012.

Late March

- Small flocks of Tree Swallows, the first true insect eaters to return to central and eastern Ontario, can sometimes be seen flying low over ice-free lakes and rivers as they search for midges. In the Kawarthas, the vanguard of a half-dozen or so Tree Swallows usually turns up over the Otonabee River.

- Sandhill Cranes return to many parts of central and eastern Ontario and can sometimes be heard calling at dawn and dusk and seen performing their courtship dance. This includes head bobbing, bowing, and leaping into the air. This species has been steadily increasing its numbers and range over the past 20 years and has now successfully colonized much of central and eastern Ontario.

The male Red-winged Blackbird arrives before the female in the spring.
Karl Egressy

BIRDSONG RETURNS IN FORCE

The dreary silence of winter has definitely lost its grip by early March. Although not the full medley of April and May, there is now a wide selection of birdsong to be heard, especially on bright, mild mornings. The songs of Northern Cardinals, Mourning Doves, Song Sparrows, House Finches, Common Grackles, Brown-headed Cowbirds, Red-winged Blackbirds, and American Robins, as well as soon-to-depart Dark-eyed Juncos and American Tree Sparrows, announce the debut of another season of avian romance. When you hear a bird singing — in nearly all cases, the male of the species — remember that he is either saying: "I'm a healthy male looking for female with which to raise a family," or "This piece of real estate and the female residing here are already taken, so back-off, buddy!" Singing is much more energy efficient than fighting or chasing away intruders. When I hear the right birds at the right time in the expected places then I know that the natural world is operating as it should and that, despite the myriad obstacles of migration, the birds of spring have once again returned.

There is a great deal of satisfaction to be derived — and frustration to be avoided — by learning at least the common songs. As Bernd Heinrich writes in *A Year in the Maine Woods*, "to walk in the woods and not recognize the songs is to not hear them."[1] Even before the leaves come out you will probably hear three or four times as many birds as you will see; with practice, they can all be identified by song. For most people, the easiest way to remember birdsongs is by using a mnemonic or memory-aid. The mnemonics that work best for me are the English "translation" variety. One of the best known of these is the "teacher, teacher, teacher" of the Ovenbird.

There is no doubt that some species sound similar to others. However, when you take into consideration the context of the song — habitat, time of year, and the bird's behaviour — the choice usually comes down to one species. The context is the secret trick that birders use to make what might otherwise seem like extraordinary acoustic identifications. Start by learning the songs of the species that are most common in your own backyard. I find that digital recordings are very helpful and easy to listen to on your computer, in the car, or on a smart phone or MP3 player. A number of excellent birdsong apps, such as the Sibley eGuide to the Birds of North America, are now available.

THE WATERFOWL SPECTACLE BEGINS

One of the highlights of March is the large number of northward-bound waterfowl that congregate on open bays, rivers, and lakes. The birds use these areas as feeding and staging grounds before pushing farther north. For naturalists and bird enthusiasts, viewing migratory waterfowl in late March and early April has always been a rite of spring. Not only are the birds in immaculate breeding plumage, but they are often quite close to shore and therefore easy to observe. Some of the ducks that are present in the largest numbers include mergansers, goldeneyes, scaups, Ring-necked Ducks, Buffleheads, scoters, and Long-tailed Ducks. You may wish to visit Presqu'ile Provincial Park near Brighton for their annual Waterfowl Viewing Weekends. They take place over two weekends in March. Some years, more than 10,000 waterfowl are present, including small numbers of Tundra Swans. Several duck species that occur uncommonly inland from the Great Lakes can be seen in abundance at Presqu'ile. These include Canvasback, Redhead, and Greater Scaup. Spotting scopes are set up for use by the public, and experts are on hand to answer questions and to help with identification. Amherst Island on Lake Ontario, Shirley's Bay on the Ottawa River, the Upper Canada Migratory Bird Sanctuary on the St. Lawrence River, and Cabot Head on the Bruce Peninsula also offer excellent waterfowl viewing in the early spring.

MAMMALS

MARCH HIGHLIGHTS

All Month Long

- Male eastern chipmunks emerge from their cozy dens and venture out into yet snow-covered woodlands and suburban backyards to find a mate.

- Male muskrats range far and wide looking for love. They will attempt to mate with as many females as they can find, which often results in vicious fights with other males. Muskrats are often found dead on the road in the spring, when their wanderings bring them into the path of automobiles.

- Watch for the meandering tracks of male raccoons in the snow and mud as they search out receptive females. They may also visit your yard in search of food and are even known to enjoy birdseed.

- March is also the mating season for other mammal species with a short gestation period. Among these are groundhogs, eastern cottontails, snowshoe hares, European hares, red squirrels, and both northern and southern flying squirrels.

- Eastern cottontail rabbits can put on quite a show at this time of year. Usually after dark, but sometimes in the early morning twilight, mating pairs of cottontails will square off in a manner reminiscent of bighorn sheep. The buck chases the doe rabbit until she eventually turns and faces him. She then actually spars at him with her forepaws. The two will then run headlong at each other as if to butt heads. However, at the last instant, one animal will jump half a metre in the air while the other runs beneath it.

- To improve your chances of actually seeing mammals, try sitting quietly in your car for 15 or 20 minutes at dawn in an area where two or more habitat types come together. This may be where a woodlot borders on a marsh or a field is bisected by a shrub or fence row. Use binoculars to scan from your car.

Early March

- Red squirrels become aggressive toward their own kind as the mating season begins. Listen for the squirrels' scolding *cherr* call as feuding over territory becomes common. Watch, too, for high-speed chases as males pursue females through the treetops.

Mid-March

- When the sap starts to run, red squirrels can sometimes be seen biting sugar maple twigs and branches. It turns out that they are actually harvesting the sap by starting a flow, waiting a day or so for some of the water to evaporate, and then returning to lap up the sugar-rich liquid. Watch for "sap icicles" at the end of branches.

Late March

- Red squirrels, gray squirrels, and red foxes bear their young any time between late March and the end of April.

- The weasel's coat may be already turning from white back to brown. The change is caused by hormones that are released as a result of the lengthening hours of daylight.

- Moose suffer from tick infestations in the winter. Winter tick (*Dermacentor albipictus*) larvae burrow under the skin, where they consume the animals' fat and cause tremendous irritation. In an attempt to get rid of the ticks, they rub, scratch and bite their hide, causing major hair loss. Some years, large numbers of moose end up dying from hypothermia in late winter and spring.

- Groundhogs finally emerge from hibernation. Anecdotal reports seem to suggest a decline in groundhog numbers in central and eastern Ontario.

Whether or not red squirrels breed successfully depends on the size of that year's seed crop.

Terry Carpenter

AMPHIBIANS AND REPTILES

MARCH HIGHLIGHTS

Late March

- If the weather has been exceptionally mild, turtles emerge from hibernation, eastern garter snakes become active, and the first spring peepers, Midland chorus frogs, and wood frogs begin calling. This is a good time to learn the calls of central and eastern Ontario's nine frogs and one toad. They are all quite distinctive and easy to remember.

Call Descriptions of the Amphibians of Central and Eastern Ontario

By using the memory aids listed below, along with a CD or online recording of their calls, you should have no trouble learning to identify our frogs and one toad species by voice. An excellent website is Nature Watch (*www.naturewatch.ca*).

Common Name	Memory Aid
American toad	Long, high-pitched musical trill lasting up to 30 seconds.
Gray tree frog	Musical, slow, bird-like trill delivered in two- or three-second bursts, almost like a "gentle machine gun."
Spring peeper	Short, loud *peep*, repeated once a second; not unlike a high-pitched toy horn.
(Midland or western) chorus frog	Short, trill-like *cr-r-e-e-e-k* with strongly rolled Rs, sounding like a thumb drawn along the teeth of a comb; repeated every few seconds.
Wood frog	Short, subtle chuckle, like ducks quacking in the distance. The call has the same grating quality as that of the leopard frog but is produced in much shorter phrases.
Leopard frog	A rattling snore followed by guttural *chuck-chuck-chuck*. Some parts of the call sound like wet hands rubbing a balloon.
Pickerel frog	Low-pitched, drawn-out snore, increasing in loudness over a couple of seconds. It has little carrying power.

Green frog   Short, throaty *gunk* or *boink*, like the pluck of a loose banjo string; usually given as a single note.

Mink frog   Rapid, muffled *cut-cut-cut*, like a hammer striking wood. A full chorus sounds like "horses hooves on cobblestones."

Bullfrog   Deep, loud, two- or three-syllable *rr-uum* or *jug-o-rum*.

Hints for remembering some of the common frog calls.
Kim Caldwell

Fish

March Highlights

All Month Long

- Winter-kill may occur in shallow lakes when the ice stays late and there is deep snow cover. Under these conditions, insufficient sunlight is able to penetrate the ice and snow to allow photosynthesis. Most of what oxygen is available is used up by bacteria to break down decaying vegetation. Large numbers of fish can sometimes die of asphyxiation.

- For many fish, activity levels begin to increase following the period of winter dormancy.

- Walleye and Northern Pike begin staging near spawning areas. Movement to these areas is triggered in part by increased flow rates as a result of snowmelt in late winter and early spring.

Insects and Other Invertebrates

March Highlights

All Month Long

- A less-than-welcome sign of spring is the reappearance of house flies. Although they lay eggs in the fall that will hatch later in the spring, many adults also survive the winter by sleeping in a hibernation-like state called diapause. Adults seen in March are individuals that have been aroused from diapause by the warmer temperatures. Watch, too, for cluster flies buzzing about on windowpanes as they try to get outside. (See October Invertebrates)

Late March

- The first swarms of mating midges can be seen over conifers, often along the shores of lakes and rivers. They will continue to be fairly common throughout the spring, summer, and fall. (See page 72)

- On a warm day in late March, you may get a glimpse of a mourning cloak butterfly taking its first flight since the previous fall. These purple-black and yellow butterflies will often feed on sap dripping from maple trees in spring. Like all anglewing butterflies, the mourning cloak overwinters in the adult stage of the life cycle.

- Warm weather may also inspire the odd honey bee to visit a blooming crocus or snowdrop in your garden. Honey bees are one of the few insects that remain active all winter.

- A sugar bush can be a great place to observe insect activity in March. Flowing maple sap attracts insects such as bees, ladybird beetles, mourning cloak butterflies, and three-spotted sallow moths *(Eupsilia tristigmata)*, a species belonging to the Noctuidae family. Like anglewings, some Noctuidae moths overwinter as adults.

- Pregnant female monarch butterflies begin leaving their winter sanctuaries in the mountains of Mexico and are streaming northward. Among these are the same butterflies that flew south from central and eastern Ontario last September! They will fan out across northern Mexico, Texas, and Louisiana, laying eggs as they go. After depositing their last eggs, the females die, having lived an extraordinary life — especially for an insect. The next generation continues the migration northward.

Mourning cloak butterflies are able to overwinter as adults, thanks to "antifreeze" proteins in the blood.

D. Gordon E. Robertson, Wikimedia

SWARMING MIDGES

Looking out over a lake or river on a sunny spring day will often reveal large numbers of tiny, dark insects flying about. These harmless, mosquito-like flies are midges (family Chironomidae). Midges are mostly aquatic although a few live in soils. Aquatic midges spend most of their lives as larvae living in the mud bottom of lakes and streams. However, after transforming into a pupa, they rise to the water's surface, where the adult midge emerges from the pupal case and flies off.

If you turn your attention to the shoreline, you may also see dense clouds of midges swarming over the top of prominent objects such as trees. These swarms are a behavioural adaptation for finding mates. The swarm itself is composed almost entirely of males who are awaiting the arrival of females. The males have large, feather-like antennae which can detect the sound of an approaching female. The female produces an "erotic" flight buzz that attains just the right pitch when she reaches sexual maturity. When the female flies into the swarm, the males move toward her like iron shavings to a magnet. One male will attach himself to her and mate. The female will then leave the horde to lay her eggs.

Take time to observe the movements of the swarm. As long as there is no wind, the insect cloud remains more or less at the same height. When a breeze comes up, however, the swarm moves lower but soon rises again as the wind subsides. These movements often suggest a plume of smoke or the synchronized wheeling of a tightly-knit flock of shorebirds or starlings. The swarm seems to function as if controlled by a single brain. There is no need to worry about being bitten because midges do not bite. In fact, the adults of many species do not appear to feed at all; their life expectancy is only a day or two.

Given their abundance, midges are hugely important in both aquatic and terrestrial food chains, even in early spring and late fall. Unlike most other insects, some midge species have the ability to fly at very low temperatures, often just a few degrees above freezing. Early migrants such as Tree Swallows are especially grateful for the cold weather flying ability of midges. The swallows feed heavily on these hardy insects as they emerge from the water almost immediately after ice-out.

PLANTS AND FUNGI

MARCH HIGHLIGHTS

All Month Long

- Depending on the species, willow twigs seem to acquire a more exuberant honey-bronze, wine-red, or mustard-yellow. Red osier dogwoods appear flushed with a more lively red. Even if the colour change is an illusion to winter-weary eyes, it is still welcome!

- Lilac, red maple, silver maple, and red-berried elder buds swell this month and stand out clearly against the blue spring sky.

Early March

- The furry catkins of pussy willows and their cousins, the aspens, poke through bud scales and become a time-honoured sign of spring. (See page 75)

Mid-March

- Sugar maple trees are tapped around the middle of March to make Ontario's world famous maple syrup. Daytime temperatures above 5°C combined with nights below freezing create the best conditions for a good sap run. (See page 75)

Late March

- Coltsfoot, with its dandelion-like flowers, may bloom if the weather is warm enough.

- Skunk cabbage blooms in March. The flowers are in a ball-shaped cluster inside a green and purple hood. The foul, skunk-like smell actually attracts early bees and insects for the purpose of pollination. This species, however, is rare in most of central and eastern Ontario, but may spread northward with climate change.

- Wild leek leaves poking through patches of late-March snow are often the first sign of new herbaceous plant growth. Leeks do not flower, however, until early summer. By this time, the leaves will have died and fallen off since the heavy shade of the forest will have made photosynthesis impossible.

- This is one time of year when dandelions, a non-native species, are actually a welcome sight. Watch for them along the bottom of south-facing walls. Dandelion leaves remain green all winter, thereby allowing the plant to respond quickly to the warmth and sunlight of early spring.

- Silver maple and red maple flowers may begin to appear.

FIRST FLOWERS

Although subtle in nature, change is afoot in our tree and shrub communities even in March. By mid-month, the pussy willow's bud scales open up, a traditional sign of spring. The pussy willow is a native shrub and one of about 15 to 20 different species of willows that can be found growing wild in central and eastern Ontario. Almost all of our native willows are shrub-like in size. Poplars and aspens are also members of the willow family, and their buds, too, are starting to open and can initially look quite similar to those of willows.

The pussy-like buds of willows and aspens are actually clusters of tiny flowers densely covered with silky white hairs when immature. *Catkin* is a Dutch word meaning "little kitten," obviously a reference to the flower's soft, kitten-like appearance. You can think of a catkin as a cob of corn in which each flower is like one of the kernels. Using this analogy, a furry pussy willow bud is essentially a dense "corn cob" of female or male flowers with each individual flower sporting its own attending silvery hairs. Seen all together, they form the attractive, fuzzy pussy willow bud we all know. The hairs serve to keep the reproductive structures warm and hasten their development by trapping heat from the sun.

Like all willows, poplars, and aspens, the male and female flowers actually appear on different trees. In other words, some pussy willow trees are male and others are female. The buds on the male pussy willow trees are showier. As spring advances, the tiny staminate flowers of the male catkins grow longer and begin to emerge through the surrounding hairs. The catkins are said to be fully "open" when the yellow pollen-bearing anthers stick out and the stigmas can be seen. On the female trees, the pistils, too, begin protruding. They are stubby looking and greenish in colour. The male flowers release their pollen just when the female flowers are prepared to receive it. As a general rule, trees that have catkins or other "non-attractive" flowers — as opposed to typical flowers with bright petals such as cherry or apple trees — are wind-pollinated. Willows, however, are also able to employ insects for the task of pollination, despite the fact that their flowers lack petals and are anything but showy. This makes them somewhat of an exception to the rule in the plant kingdom. That being said, willows do have sweet-scented pollen and nectar to attract pollinators. Early flowering confers a certain advantage to the willows. When willow flowers mature in April or early May and are ready for pollination, there is relatively little competition with other plants for pollinating insects.

Pollen-bearing male pussy willow flowers as they appear later in the spring.
Terry Carpenter

AT THE SUGAR BUSH

The annual gathering of the sweet sap of the sugar maple has always been a powerful symbol of this time of transition between winter and spring. Blanketed in snow when trees are tapped, the forest floor will have become a bare carpet of brown when the sap flow finally stops. As with all trees, the sap of the sugar maple is a mixture of water and minerals taken up from the ground and sugar that has been stored in the wood. Although sap is present in the tree all year long, it is only sweet enough to make into syrup for a few weeks each spring. To understand the whole process, we have to go back to the previous summer. Chlorophyll, the magical green pigment of the leaves, uses energy from the sun to convert carbon dioxide and water into sugar and oxygen through the process of photosynthesis. The sugar is converted into starch and serves as the tree's food reserve to fuel growth. Excess starch is stored in the wood and, over the course of the winter, converted to sucrose. As spring approaches, the sucrose dissolves into the sap to be used to fuel the tree's spring growth. A good sap year depends to a large extent on the growing conditions the summer before. The sugar content of sugar maple sap is generally 2 to 3 percent but can be as high as 7. Although other maples, such as the red and the silver, can also be tapped, their sap is not nearly as sweet, nor does it have as pleasing a flavour.

In order to have strong sap flows, a suitable temperature cycle of warm days (2° to 7°C) and cool nights (-4° to -6°C) must develop. When the days are warm, pressure in the trees increases. This causes a sap flow from all directions — not just from the roots up but also from the upper branches down. As long as the tree's internal pressure is greater than the atmospheric pressure outside, the sap will move. The sap flows through an area of the outer tree trunk called sapwood, where actively growing cells conduct water and minerals from the roots to the branches. In order to gather the sap, a hole about eight centimetres long is made in sapwood and a metal tap is inserted. The hole can be thought of as a "leak" in the tree to which sap flows. At night, when the temperature drops to below freezing, the tree's internal pressure becomes less that the surrounding air pressure. Suction develops and water and minerals are sucked up into the tree by the roots. The tree is essentially recharging itself, allowing for sap to flow during the next warm period. It generally takes 30 to 40 litres of sap to produce one litre of syrup. As soon as the buds begin to expand and open, the sap becomes off-flavoured or "buddy" and is no longer gathered.

"FIRST BLOOM" CALENDAR FOR SELECTED TREES, SHRUBS, AND HERBACEOUS PLANTS

For each of the spring and summer months, the "first bloom" calendar gives the approximate time when some of the better-known species begin to flower. The year-to-year variation in weather, however, may accelerate or slow the flowering process.

Mid-March: Pussy willow
Late March: Silver maple, red maple, poplars, and aspens

WEATHER

MARCH HIGHLIGHTS

All Month Long

- We begin the month with about 11 hours of daylight and end with over 12 ½ — a gain of more than 90 minutes!

- There are few rules when it comes to March; most any type of winter or spring weather can be expected.

- For many of us, the characteristic smell of March is that of making maple syrup — the sweet aroma of evaporating sap, mixed with the smell of wood smoke, wafting from the sugar shack. The familiar smell of thawing earth and organic matter is usually everywhere by month's end, as well.

- The sun's heat melts the snow surface during the day so that it freezes into a hard crust at night. Some years it becomes hard enough to walk on.

- Dark objects such as tree trunks and fallen twigs and leaves absorb sunlight and transform it into heat. This heat melts the surrounding snow, allowing the object to sink down.

- Melting snow uses up heat from the surrounding air, making the air feel cooler than you might expect.

Early March

- When we move our clocks ahead one hour on the second Sunday in March, sunrise is suddenly an hour later than the day before. This means getting up in the dark once again if you're an early riser.

Mid-March

- On or about March 18, day and night are almost exactly equal in duration.

Late March

- The spring equinox falls on or about March 21. Both the moon and sun rise due east and set due west that day. If you need a reason to celebrate, we are gaining daylight faster now than at any other time of year — over two minutes a day! For the next six months, we can enjoy days that are longer than nights.

- The snow cover is usually gone by late March in most areas of central and eastern Ontario, but occasional snow flurries are still to be expected.

MARCH WEATHER AVERAGES, 1971–2000[2]

City or Town	Daily Max. (°C)	Daily Min. (°C)	Rainfall (mm)	Snowfall (cm)	Precipitation (mm)
Owen Sound	3	-5.4	39.8	37.6	77.4
Huntsville	2.6	-8.9	35	34.4	69.5
Barrie	3.2	-7.5	28.9	28.1	57
Haliburton	2.4	-9.5	43.7	33.2	76.8
Peterborough	3.4	-6.3	42.1	26.6	67.9
Kingston	2.8	-5.9	47.5	29.9	79.5
Ottawa	2.1	-7	33.6	32.1	64.9

APPROXIMATE MARCH SUNRISE AND SUNSET TIMES (EST/DST)[3]

(Note: Twilight starts about 30 minutes before sunrise and continues about 30 minutes after sunset. Daylight Savings Time begins second Sunday in March.)

Location	Date	Sunrise (a.m.)	Sunset (p.m.)
Grey-Bruce	Mar. 1	7:03	6:10
	Mar. 15	7:38	7:28
	Mar. 21 (equinox)	7:26	7:36
	Mar. 31	7:08	7:48
Muskoka/Haliburton	Mar. 1	6:57	6:05
	Mar. 15	7:33	7:23
	Mar. 21 (equinox)	7:22	7:31
	Mar. 31	7:03	7:43
Kawartha Lakes	Mar. 1	6:52	6:00
	Mar. 15	7:27	7:18
	Mar. 21 (equinox)	7:16	7:25
	Mar. 31	6:58	7:38
Kingston/Ottawa	Mar. 1	6:44	5:53
	Mar. 15	7:20	7:11
	Mar. 21 (equinox)	7:09	7:18
	Mar. 31	6:50	7:30

NIGHT SKY

MARCH HIGHLIGHTS

All Month Long

- Major constellations and stars visible (March 15 — 8:00 p.m. EST)

  Northwest: Andromeda (with M31 galaxy) is setting; Cassiopeia just above to the right; Pleiades above to the left; Taurus (with *Aldebaran*) due west.

  Northeast: Bootes (with *Arcturus*) low in sky; Virgo (with *Spica*) just above east horizon; Big Dipper standing upright fairly high; Little Dipper (with *Polaris*) to its left.

  Southeast: Leo (with *Regulus*) rules over the southeast sky.

  Southwest: dominated by Orion (with *Betelgeuse* and *Rigel*); Taurus (with *Aldebaran*) high to its right; Canis Major (with *Sirius*) low to its left; Canis Minor (with *Procyon*) due south; Auriga (with *Capella*) high above Orion; Gemini (with *Castor* and *Pollux*) to left of Auriga near zenith.

- The Algonquian name for the full moon of March is the Worm or Maple Sugar Moon.

- No other season offers as many bright stars and constellations as spring. There are no less than 11 first magnitude stars visible. (See page 79)

- Starting at about 7:30 p.m., watch the northeastern horizon for Arcturus, the brightest star of summer and the harbinger of spring. It is also the alpha star of the constellation Bootes. To find this beautiful star, follow the arc of the Big Dipper's handle to the next brightest star. Mariners used to say, "arc to Arcturus."

- The signature constellation of spring is Leo. Although somewhat dim, it is one of the few constellations that actually looks like its namesake. Leo is best imagined as a lion sitting on its haunches.

- At the spring equinox, both the moon and sun rise due east and set due west.

- Around the time of the full moon closest to the spring equinox, the moon rises later each subsequent night than at any other time of year. For example, there is about an 80-minute difference in the time of moonrise between the evening of the full moon and the evening after. This is just the opposite of what happens at the fall equinox during the Harvest Moon when the moon rises an average of only 25 minutes later from one night to the next.

BRIGHT SPRING STARS!

No other season offers as many bright stars and constellations as spring. There are no less than 11 first magnitude stars visible. Two of the most dominant stars of the spring sky are Arcturus, in the constellation Bootes, and Spica, in the constellation Virgo. Arcturus is a yellow-orange star and, among stars visible from central and eastern Ontario latitudes, is second only to Sirius in brightness. Spica is a bluish star and not quite as bright. The Big Dipper, which is now almost at the zenith, can help you locate both of these stars. Follow the curve of its handle (going away from the bowl) and "arc to Arcturus" and then continue the arc or "speed on to Spica." By following the curve of the Big Dipper's handle in the other direction, it is easy to locate Regulus, the brightest star of the Leo constellation. If you look to the southwest, the winter stars such as Betelgeuse, Rigel, Sirius, Procyon, Aldebaran, Capella, Castor, and Pollux are still very prominent. And, by staying up after midnight, you can even get an early glimpse of the upcoming summer constellations.

In earlier times, it was often the custom to plant or harvest crops when a particular star first appeared above the eastern horizon just before sunrise. In some cultures, the first appearance of Spica was the signal that the planting of wheat should begin.

APRIL

Frog Song and Sky Dancers

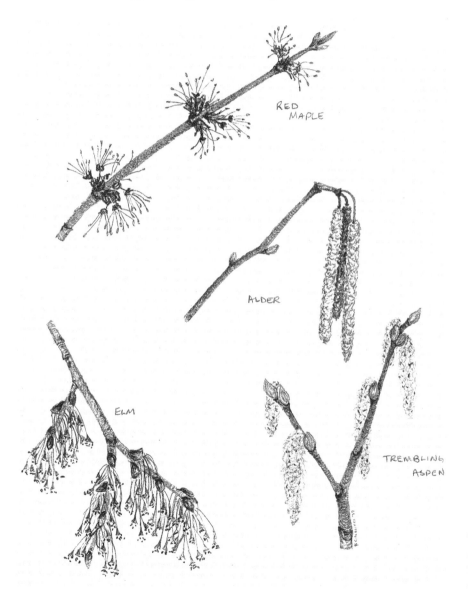

RED MAPLE

ALDER

ELM

TREMBLING ASPEN

The flowers of several early-flowering trees.
Kim Caldwell

Every spring is the only spring — a perpetual astonishment.

— Ellis Peters

April is the time of amphibian love, when marshes, swamps, and woodland ponds reverberate with the calls of countless frogs consumed by a mating frenzy. Salamanders, too, join the fray as they venture over ice, rock, and road to make their way back to ancestral breeding ponds. High overhead, a snipe flies in wide, reckless circles, its wings creating a haunting tremolo sound. From a nearby thicket, the nasal *beep* of the woodcock is constantly repeated until the bird suddenly launches itself into the air and begins its spectacular sky dance.

On an April morning, the chorus of robins, cardinals, and Mourning Doves is so loud that you have to get up and close the bedroom window. Forests resonate with the drumming of grouse and the courtship hammering of woodpeckers. Evening Grosbeaks call from overhead as they search out swelling buds on which to feast. On our lakes, we hear other April music — the tinkling of black candle ice, the clamour of ice piling up in ridges, and the roar of waves rolling under the disintegrating frozen surface. Walk through the forest and your nose will recognize the time of year by the smell of the sodden, thawing earth and decaying leaves. And, for those of us old enough to remember, April will always be synonymous with the smell of grass fires.

New plant life, too, vies for our attention this month. The yellow flowers of coltsfoot push forth among roadside stones and debris. Almost overnight, the treetops appear less open as dormant buds swell and thicken. The flowers of maples, poplars, elms, and alder stand out against the grey-brown landscape and provide a foretaste of what is to come; because, more than anything, April is a time of great expectation. In only a few weeks, the explosive growth of buds, flowers, shoots, and leaves will totally transform the landscape. The extraordinary surge of life that we see and feel everywhere frees us of our late winter blahs and whets our appetite for the pleasures of May.

Birds

April Highlights

All Month Long

- One of the most wonderful features of early spring is that birdsong has once again returned in full force to the natural world. (See page 85)

- The courtship flight of the American Woodcock provides nightly entertainment in damp, open field habitats. (See page 86)

- Close to 30 species of birds are already nesting this month. Among these are the Osprey, American Crow, Common Grackle, Blue Jay, Eastern Bluebird, and the American Robin. The female robin — the one with the dull orange breast — selects the nest site and does most of the nest building. (See page 87)

- The pace of migration accelerates noticeably with the arrival of wading birds, Barn Swallows, sparrows, kinglets and, by month's end, the first warblers.

- The muffled drumming of the Ruffed Grouse is one of the most characteristic sounds of April. It is often described as sounding like an old truck engine slowly starting up.

- Woodpeckers are also very noisy this month. Listen especially for the courtship drumming of one of our two migratory woodpeckers, the Yellow-bellied Sapsucker. The sapsucker often hammers on dead, resonant tree trunks or on metal surfaces such as drain pipes.

- Half-crazed cardinals, robins, and even woodpeckers sometimes peck aggressively at windows and car mirrors. Being very territorial birds, they instinctively attack other individuals of the same species — in this case, their reflected image — in an attempt to drive the "invader" out of their nesting territory. Both males and females are known to do this. The banging can start at dawn and last until dusk. It may even go on for several weeks. The simplest solution to this problem is to tape a piece of cardboard over the section of the outside of the window where the bird is pecking. Hormonal changes will eventually make the birds turn their attention to other matters.

- Cold, wet springs can cause nest failures, especially in sensitive species such as Tree Swallows and Eastern Bluebirds. However, thanks to the success of well-designed bluebird boxes, bluebirds are once again relatively common in central and eastern Ontario.

- April is a very busy time for feeders. Northward-bound American Tree Sparrows and Dark-eyed Juncos move through in large numbers. Watch also for the large, thrush-like Fox Sparrow. Some years, Pine Siskins and Common Redpolls will linger at feeders in central and eastern Ontario for several weeks as they move northward toward nesting grounds.

- Common Loons return and take up residence on our lakes as soon as the ice goes out.

Early April

- Migrating waterfowl continue to move northward, with numbers usually peaking early in the month. However, depending on the amount of ice cover farther north, a great deal of variability in water bird concentrations is usually noted from one year to the next. Check out large lakes and rivers for species such as Ring-necked Ducks, Buffleheads, Common Goldeneyes, mergansers, and scaup. Flooded cornfields are also worth investigating.

Mid-April

- Up until the late 1990s, there were usually several days in mid-April when thousands of migrating Tree Swallows could be seen flying over water bodies such as the Trent-Severn Waterway and adjacent roads and fields. Swallow populations of all species have declined markedly, however, especially inland from the Great Lake shorelines.[1]

- Hundreds of Red-necked Grebes gather on the open waters of Georgian Bay between Dyer's Bay and Cabot Head on the Bruce Peninsula. They are joined by Horned Grebes, Common Loons, and deepwater ducks such as Long-tailed Ducks and scoters.

- Small flocks of Greater and Lesser Yellowlegs, a type of shorebird with bright yellow legs, can often be seen feeding in flooded fields and around the margins of ponds.

Late April

- After a winter in the southeastern United States, flocks of Yellow-rumped Warblers return to central and eastern Ontario. The male is quite stunning in his blue-grey back feathers, black breast, and yellow patches on the rump, sides, and crown. Listen for a loud *check* call note.

A Yellow-bellied Sapsucker drumming on a drain pipe to advertise his territory.
Kim Caldwell

- White-throated Sparrows are passing through and are easily attracted to feeders, as long as you put seed on the ground. They are also a great entrée into the world of birdsong, since their plaintive, whistled "Oh-sweet-Canada-Canada-Canada" song is one of the easiest to learn.

- The Ruby-crowned Kinglet is a common but little-known April migrant. Smaller than a chickadee, this hyperactive grey bird has a distinctive white eye ring and a long, boisterous call that would seem to come from a much larger bird.

- Northern Canada Geese (*Branta canadensis interiori*) bound for the Hudson Bay and Arctic Lowlands pass over in late April and early May and are a time-honoured sign of spring. They can be distinguished from our local "Giant" Canada Geese by the larger numbers in the flock and the higher altitude at which they usually fly. Giant Canada Geese (*Branta canadensis moffati*) were reintroduced into southern Ontario in the late 1960s after coming close to extinction.

THE BIRD MUSIC OF APRIL

By early April, it's hard not to be aware of the abundant birdsong flooding the airwaves. Even in suburban areas, the early morning cacophony of song often puts an end to any plans you may have to sleep in. Although the multitude of sounds may at first seem terribly confusing, almost all of the voices belong to fewer than a dozen species, at least in towns and cities. The dominant singer in built-up areas is quite often the Northern Cardinal. Cardinals sing loud, lusty, one- or two-note phrases over and over again. One of the most common cardinal songs starts with a series of drawn-out *tweer-tweer-tweer-tweer* whistles, followed immediately by a rapid string of *whit-whit-whit-whit* notes. Competing for supremacy over the sound waves but singing a little more softly, the voice of American Robin is almost impossible to miss. The song is a series of low, whistled phrases, often described in English translation as "cheerily-cheer-up-cheerily." Robin's are especially vocal at sunrise.

Three other players, albeit of lesser musical virtuosity, are also quite dominant most days. The raspy, nasal caw of the American Crow is an almost constant presence. Its close cousin, the Blue Jay, is usually heard, too, blasting out a harsh, raucous *jaay-jaay-jaay*. The harsh *chack* of Common Grackles are generally part of the cacophony, as well. These birds often fly about in early spring in small, noisy flocks.

Once you've sorted out the vocal bullies of the morning chorus, it's time to concentrate on two of the softer voices. These almost always include the soft, clear, two- or three-note whistle of the Black-capped Chickadee. Some people remember it as "Hi sweetie." You should also listen for a very high-pitched, musical series of warbled notes that are delivered with machine-gun rapidity. These belong to the House Finch.

If you happen to live in a more rural area that is interspersed with fields, the call of the Killdeer can be a common sound. This member of the plover family makes a sharp, piercing *kill-deer-kill-deer-kill-deer* that is often given in flight. This same habitat is also home to the Eastern Meadowlark. It delivers a beautiful, clear whistle that can be heard at considerable distances. The mnemonic for this song is "spring of the year." Damp fields and wetlands have several other species to offer. Most noticeable is the loud, strident *conk-a-reeeeeeee* of the Red-winged Blackbird. They also produce high, clear whistles that serve as alarm calls.

SKY DANCERS

One species that deserves some special attention in April is the American Woodcock. Its courtship flight has even earned it the name of "sky dancer." From early April to late May, the nasal *peent* of the woodcock is a common sound in damp, open habitats bordered by second-growth forest or wetland. The peenting begins in the twilight period after sunset. As darkness falls, the calls become more numerous until the bird suddenly launches itself into the air and climbs in wide circles to an altitude of about 100 metres. Because the woodcock's three outer wing feathers are extremely stiff and narrow and spread apart during flight, the air rushing through causes them to vibrate, producing a high twittering sound. The twittering stops, however, as the bird begins its zigzag descent. The sound now changes to liquid, warbled notes that the wood-cock actually sings. The warbling grows louder and louder as the bird approaches the ground, but then ends abruptly; the final portion of the descent is silent. The woodcock usually lands close to the same spot from where it took off. The bird then walks stiff-legged in the direction of the nearby female and begins the peenting sound again. A few minutes later, the poor woodcock — which must be close to exhaustion — launches into yet another flight.

Woodcock-watching on a calm April night, often accompanied by spring peepers calling in the background, is an event not to be missed. Because the darkening sky makes it difficult to actually see the woodcock in flight, try to face west so that the bird will stand out against the lighter western sky. After it takes off, you can move closer to the takeoff point. By remaining quiet and staying low, it is sometimes possible to get a close look at the woodcock when it lands. The nuptial flights usually stop when darkness falls, although some woodcock will display even during the night when there is a full moon. The birds will also display again at dawn. Woodcocks usually continue their nuptial flights until late May, well after mating has taken place. Almost any damp field overgrown with scattered shrubs can be home to American Woodcock.

If the area is wet enough, you may also hear the courtship flight of another sky dancer, the Wilson's Snipe. Actually seeing the bird can prove more difficult, because it never seems to be where the sound is coming from. The male snipe flies in wide, horizontal circles above the ground and regularly drops or dives, its tail fanned. With each dive, the outer tail feathers vibrate and make a strange tremolo sound known as winnowing. The common origin of the woodcock and the snipe is evident in the similarities in their displays; snipe, however, court actively all day long, not just at dusk and dawn.

The American Woodcock puts on an amazing aerial courtship display in spring.
Karl Egressy

NESTING ROBINS

One of the reasons we have a special fondness for robins is because they are one of the few birds to treat us to intimate views of their family lives. The nest is built by the female over the course of about six days. The basic technique involved is to drop grass on top of a layer of mud and to mould it into shape by sitting, squirming, pushing with the wrist of the wings, and stamping with the feet. Female robins seem to have an all-consuming impulse to build a new nest for each brood, so you don't need to hesitate about removing old nests. The female generally lays four blue eggs, which she incubates for 12 to 14 days. The male spends most of his time bringing food to the nest for the young and occasionally for the female as well. The chicks are stuffed with high-protein earthworms six or seven times an hour. With this kind of diet, it's no wonder that they balloon to ten times their birth weight after only ten days. In addition to caring for his offspring, the male also continues to sing heartily throughout the nesting season in an effort to tell other males to stay clear of his mate.

Robin chicks instinctively know to poop immediately after swallowing food. You might wonder how the parents manage to keep the nest clean. As it turns out, the chicks produce a kind of self-generated disposable diaper. Their urine and feces emerge together in a "bag" called a fecal sac, which is made of thick, strong mucous. Because this process happens within seconds of eating, evolution has assured that the parents will still be there to dispose of the bag before the chick sits on it and soils the nest.

The young are fully feathered ten days after hatching and are ready to leave the nest after 14 days. Fledglings are still dependent upon parental care and feeding, however. The male robin more or less takes responsibility for the fledglings and will continue to feed them for up to three weeks. In the meantime, the female will usually lay another batch of eggs, most often in a new nest. If you find a fledgling robin, or any other bird species for that matter, the general rule is to leave it alone. Most baby birds on the ground are not orphans, nor are they sick or injured. If, however, you feel a fledgling would be in danger if left where it is, you can pick it up and place it on a sheltered branch, out of harm's way. Birds have a weak sense of smell, and the parents will continue to care for the chick. Sadly, many robin nests, especially in urban areas, are destroyed by predators such as crows, jays, grackles, cats, squirrels, and even chipmunks. You can still do your part, however, by always keeping your cats indoors and encouraging your neighbours to do the same.

MAMMALS

APRIL HIGHLIGHTS

All Month Long

- When our lakes and rivers begin to open up, watch for otters bringing prey up onto the ice to eat. The river otter has been expanding its range in recent years and is increasingly common in areas south of the Canadian Shield.[2]

- Many members of the weasel family are contributing to the mammalian baby boom this month. They include short- and long-tailed weasels, American mink, martens, otters, and fishers.

- Black bears emerge from hibernation this month. (See page 89)

- After their late-winter mating season, eastern cottontails are giving birth. The young, naked and blind, are usually found in a fur-lined depression under a shrub or in tall grass. They grow so fast that they are ready to live on their own after only a month. Cottontails will continue to breed until September.

Mid-April

- The first bats come out of hibernation and take flight on mild evenings. Females move to maternity roosts, where they will give birth in late May or June. The roost might be a hole in a tree, an accessible cavity in a man-made structure such as an attic, or a bat house. These colonies may house anywhere from a dozen adults to several thousand.

- Last spring's beaver cubs are driven from the parental pond and forced to wander widely in search of a new territory.

- Porcupines switch from a diet of tree bark to one of new leaves and buds. Aspen catkins and sugar maple buds are favourite foods at this time of year.

BEARS EMERGE FROM HIBERNATION

Black bears become active again this month. Male bears are the first to emerge from their winter dens, followed by barren females and females with yearling cubs. Mothers with new cubs are the last to emerge. A bear will have lost 15 to 40 percent of its fall weight by the time it comes out of hibernation and will continue to lose weight for several more months. It will not fully gain its weight back until berries become abundant in midsummer. Although the black bear may appear somewhat emaciated in the spring, its coat is surprisingly luxuriant, having grown all winter.

In April and May, bears are forced to adopt a mostly vegetarian diet. They graze heavily on grass and even eat dandelions because of their high nectar content. When the aspens leaf out, the bears also eat huge quantities of the new, tender leaves. It is not uncommon to see a mother bear with her two yearling cubs eating away right in the top of an aspen. In the spring, adult black bears also target newly born animals such as moose calves and beaver kits. Some males will even kill and eat black bear cubs at this time of year. Clearly, everything is food for a black bear.

AMPHIBIANS AND REPTILES

APRIL HIGHLIGHTS

All Month Long
- Wetlands come alive with the clamorous calls of thousands of frogs. The first voices usually heard are those of the chorus frog, spring peeper, and wood frog. (See page 90)

- On rainy, mild nights when the first frogs begin calling, salamanders are also on the move as they head to woodland ponds to breed. (See page 91)

- Turtles come out of hibernation and can be seen sunning themselves. Spring is one of the best times to see Blanding's turtles basking in the sun, especially in Shield wetlands.

- Eastern garter snakes become active once again and get right down to the business of mating. All of the male snakes in the vicinity will converge upon a receptive female in a spaghetti-like "knot of snakes." Only one lucky male, however, gets to impregnate her. He will leave behind a plug of gelatin in her cloaca, essentially a "chastity belt" that blocks copulation by other males.

Late April

- The snore-like calls of leopard frogs join the amphibian chorus already well underway. Although far less common, it is also possible to hear the first pickerel frogs.

THE AMPHIBIAN CHORUS BEGINS

Countless thousands of vociferous frogs, caught up in the act of procreation, are putting on a wonderful show this month. The clamorous calls, given only by the males, serve to attract females and, in the case of some species, to advertize ownership of territory. When a receptive female arrives, the male clasps her waist from the back and spreads his sperm onto the eggs as they are voided from her body. All of the species that breed in early spring anchor their egg masses well below the surface, where they will not become frozen in the ice on a cold night.

The frogs of early spring usually begin to call around the middle of April when nighttime air temperatures have warmed to at least 7°C. The first species to break the long silence of winter are the chorus frog, spring peeper, and wood frog. Only about two centimetres long, the chorus frog has a rising trill that sounds quite similar to a finger being drawn along the teeth of a comb. Like the chorus frog, spring peepers also come out of hibernation with the onset of the first warm rains. Amazingly, they begin singing their hearts out more or less immediately, despite having had nothing to eat for five months and having spent the winter frozen under a blanket of snow on the forest floor. The peeper's call is a surprisingly powerful series of high-pitched, piercing, bird-like peeps that are repeated about once per second. They are sometimes accelerated to form a short trill. A full chorus of peepers at close proximity is almost physically painful to the ears. I find that even hours afterward, the peeper chorus can continue to sound in my head.

In more wooded wetlands, the aptly named wood frog is also adding its contributions to the soundscape. Looking like a masked thief, this handsome frog produces a short chuckle, almost as if it were doing an imitation of ducks quacking. Finally, before the end of the month, the first leopard frogs are usually heard. Their call is often described as a rattling snore followed by a series of guttural chucks. Frog calls are loudest during the first few hours of darkness and fall off after midnight. However, chorus, wood, and leopard frogs also call a great deal during the day.

A large chorus of spring peepers is sometimes compared to the sound of sleigh bells.

Joe Crowley

SALAMANDER WATCHING

Unlike frogs or toads, salamanders do almost nothing to attract attention. However, the month of April affords us a special opportunity to observe these enigmatic creatures. At about the same time as the peepers start calling, salamanders, too, are participating in an ancient rite of spring. On mild, rainy nights, thousands of these gentle creatures are making their way over snow, ice, rock, and pavement to breed in their ancestral ponds. Running a gauntlet of skunks, raccoons, and automobiles, the salamanders are likely following an imprinted memory of their birthplace with its specific odours of mud and decaying vegetation. The three species most often seen are the spotted sala-mander (sometimes referred to as the yellow-spotted), the blue-spotted salamander, and the eastern newt. The first two species are members of the mole salamander genus, *Ambystoma*, and measure about 10 to 14 centimetres in length.

Salamander mating begins with a sort of underwater dance in which large groups of males gyrate and rub up against the females. If the female is willing, the pair will leave the group and the male will deposit a spermatophore (a small, jelly-like packet of sperm) on underwater debris. The female becomes fertilized by taking the spermatophore into her genital opening. Later, she will deposit from one to three gelatinous clumps of eggs on underwater vegetation. In about six weeks, tiny, gilled tadpole-like larvae are born, and will leave the water by summer's end. By using temporary ponds and wet areas, salamander eggs and larvae are usually spared predation by fish and turtles. However, there is a trade-off. The salamanders must breed very early in the season to assure there is enough time for the larvae to transform into air-breathing terrestrial salamanders before the water dries up in summer.

To see salamanders firsthand, wait for an evening of gentle, prolonged rain in early to mid-April when the first frogs start calling. Armed with a strong flashlight, walk or slowly drive along back roads that pass through woodlands and treed, swampy areas. By watching carefully, you should be able to see the salamanders on the road or in open areas close to nearby ponds.

Although most populations of amphibians in central and eastern Ontario still appear to be healthy, this is not necessarily the case elsewhere. Frogs, toads, and salamanders are among the most at-risk animals on earth. Of the world's 6,000 or so known amphibian species, fully one-third is threatened with extinction. In order to be able to better protect Ontario's amphibians in the future, the Ontario Reptile and Amphibian Atlas project is now underway. The project's main goal is to improve our still-incomplete knowledge of the distribution and population levels of the province's amphibians and reptiles. The Atlas needs your help. (See "An Uncertain Future for Snakes" on page 165 for details)

The spotted salamander is a beautiful black amphibian with yellow spots.
Scott Camazine, Wikimedia

AVERAGE BREEDING PERIODS FOR FROGS AND TOADS IN CENTRAL AND EASTERN ONTARIO

Common Name	Breeding Period
Spring Peeper	mid-April to late June
Chorus and wood frog	mid-April to early May
Leopard and pickerel frog	late April to late May
American toad	early May to late May
Gray tree frog and mink frog	mid-May to late June
Green frog and bullfrog	mid-May to late July

FISH

APRIL HIGHLIGHTS

All Month Long

- April is the fish-watching month par excellence. Spring is in the air and "in the water." Soon after the ice goes out, many species of fish move toward the sun-warmed shallows in search of food. A large number of species also spawn at this time.

Mid-April

- Walleye lay their eggs in rocky, fast-flowing stretches of river and along shoals in lakes. They are best seen after dark, so take a strong flashlight. At the same time or shortly after, hordes of suckers also spawn in fast-flowing waters as well as in streams. (See page 94)

- Rainbow Trout leave the Great Lakes to move upstream to spawn in the shallow riffles of streams such as those on the Oak Ridges Moraine. They are a spectacular sight as they jump up the fish ladders at locations such as the Ganaraska River in Port Hope, the Sydenham River in Owen Sound, and the Beaver River in Thornbury.

Late April

- Northern Pike spawn in flooded areas when the water is between 4° and 11°C.

- Muskellunge spawn somewhat later on emergent vegetation at water temperatures of 9° to 15°C. (See below)

- Yellow Perch, too, will reproduce in April if the water is warm enough. They spawn right after the suckers and deposit their gelatinous mass of eggs on vegetation or woody debris in the water.

NORTHERN PIKE AND MUSKELLUNGE

Northern Pike and Muskellunge spawn in the early spring shortly after the ice goes out and the water temperature is between 4° and 11°C. By mid to late April, pike move into weedy, shallow bays and flooded marshy areas to scatter and then desert their eggs. They will sometimes spawn in water so shallow that half the fish sticks out into the air! If you quietly walk or canoe along a marshy shoreline or flooded area, you may see what looks like a sleek, miniature submarine cruising along the edge of a cattail bed or between grassy hummocks. This is almost certainly a pike searching for a suitable weed bed on which to scatter eggs or sperm.

A short time later, when the water has warmed to between 9°C and 15°C, Muskellunge can be seen spawning in similar habitat. Muskies prefer to spawn along the edge of beds of emergent plants such as cattails. The eggs are deposited over stumps and various types of vegetation by large females, usually accompanied by two or more smaller males. It is quite entertaining to watch the rapid swimming and rolling of the spawning fish. In order to see this spectacle, check out areas of suitable habitat once the water reaches 9°C. Pike are considered a threat to Muskie populations in some areas such as the Kawartha Lakes because they spawn earlier than Muskies and prey upon them. Pike are not native to the Kawarthas and pose a serious and possibly lethal threat to the unique strains of Muskellunge which have inhabited these lakes for thousands of years.[3]

WALLEYE AND WHITE SUCKERS

Starting sometime around the middle of April, when water temperatures reach 5°C, Walleye begin to spawn. There are two different spawning modes. Some populations breed exclusively in fast-flowing water over gravelly and rocky bottoms. Other fish spawn on rocky, wind-exposed shoals in larger lakes. After the eggs are broadcast by the female and fertilized by the male, they are deserted. Because the growth of the young is based on water temperature, the best reproductive success occurs in years when the Walleye spawn a little later than usual, and there is a steady warming of water temperature without any significant cold periods. Extremely cold water or dramatic fluctuations in water temperature can kill the fry. The spawning period can last up to three weeks.

The Walleye spawning period provides an opportunity for excellent fish-watching. Although some fish can be seen during the day, most are observed only at night when the actual spawning takes place. For night viewing, you will need a strong-beamed flashlight or portable spotlight. The Walleye's eyes glow when a light is shone on them. Spawning Walleye can be seen in the fast water areas below almost any dam in the Kawarthas, such as Lock 19 in Peterborough and Lock 32 in Bobcaygeon. Other good Walleye-watching sites include the village of Westport (north of Kingston) and where Highway 7 crosses the Mississippi River in Innisville.

Shortly after Walleye have begun spawning, suckers join the fray. They move up from lakes in the spring to deposit their eggs and sperm in fast-flowing sections of rivers and streams. Sometimes the schools can be so dense that a small stream may appear to be wall-to-wall fish, as they splash and scramble by the thousands over rocky rapids. The sucker is a prolific spawner, depositing an average of 50,000 eggs. Suckers are also one of the most easily watched spring spawners and can be seen both day and night. They are a favourite spring food of bears, which have no trouble catching them in shallow water.

The large eyes of these spawning Walleye allow them to see well in low light conditions.
Matthew Garvin

INSECTS AND OTHER INVERTEBRATES

APRIL HIGHLIGHTS

All Month Long

- Early butterflies such as tortoiseshells and anglewings are sometimes seen flying, basking in the sun, or feeding at sap flows. (See page 96)

- Bees can be seen foraging at early blossoms such as crocuses. However, all is not well with bee populations. (See page 96)

- Woolly bear caterpillars become active after overwintering as larvae curled up in some protected nook or cranny. They resume eating leaves for a short while and then pupate in a cocoon made from their own hairs. About two weeks later, they emerge from the cocoon as white Isabella moths.

- Large, pregnant queen wasps and bumblebees are often seen on warm days in early spring. Bumblebees fly low to the ground in search of a hole or burrow to claim as a home. The female makes a small ball of grasses and mosses in which to deposit her eggs.

- On warm days, mosquitoes of the *Culex* genus may appear. They have overwintered as adults.

- In areas where birch and aspen grow, watch for the Infant moth (*Archiearis infans*), a small day-flying moth with orange and black hindwings.

- Clouds of strange crustaceans known as fairy shrimp (genus *Eubranchipus*) can often be seen in meltwater pools immediately after the snow and ice disappear.

Late April

- If the weather is warm, the first spring azure and elfin butterflies are seen by month's end.

- Warm weather may also see the emergence of the first dragonflies. Often the earliest species to appear is the Hudsonian whiteface. By month's end, a warm push of air from the south may also bring with it the first common green darner dragonflies of the season. This species has two distinct populations, one which is migratory and one which stays in central and eastern Ontario and overwinters as a nymph. Early darners are the offspring of those individuals that migrated south in the fall. When they arrive in the southern United States, they mate, lay eggs, and then die. The generation that emerges from these eggs makes the trip back to Ontario.

BUTTERFLIES OF EARLY SPRING

On a warm spring day when temperatures climb above 15°C, you may be surprised to see a butterfly in flight or even feeding on sap oozing from a freshly cut tree stump. In fact, several species of our butterflies overwinter in the adult stage and become active again even before the last snow melts and the first flowers bloom. Species of the genus *Nymphalis* (the tortoiseshell butterflies) and, to a lesser degree, *Polygonia* (the anglewings or commas) overwinter in central and eastern Ontario. They choose protected hideaways such as hollow trees or logs, woodpiles, and loose boards on outbuildings.

Tree sap is probably the most common source of butterfly food in early spring. They are attracted to sap oozing from recently cut stumps, from broken branches, and from tree taps at the local sugar bush. The Yellow-bellied Sapsucker is also an important ally of butterflies at this time of year. When sapsuckers return in April, they immediately begin to drill their characteristic series of horizontal, pit-like holes, often in birch and hemlock. The bird feeds both on the sap itself and on the many small insects that the sap attracts.

The most common butterfly species seen at this time of year is usually the mourning cloak. This large tortoiseshell butterfly has purple-brown wings edged with yellow. It will often feed with its wings wide open, thereby allowing a good view of the wing colour. In early spring, the sunny edges of mixed and deciduous forests as well as wooded cottage neighbourhoods are usually the best places to see butterflies. Many species such as elfins like to sit on sunny roads in order to absorb heat.

THE WONDER OF POLLINATION BY BEES — BUT FOR HOW LONG?

An understanding of bees and their pivotal ecological role is, in many ways, an understanding of how nature works. The essence of the game is that many plant species need to entice bees into pollinating them — in other words, transferring the male sex cells from the stamens of a flower on one plant to the sticky, receptive pistil, or female organ, of a flower on a different individual of the same species. This results in cross-pollination — a mixing of genetic traits between different species — which produces a new generation of plants that may be better adapted to changes in the environment. The flowers are luring the bees in with brightly coloured petals, alluring smells, and rewards of nectar and the pollen itself. The nectar is strategically located near the sexual organs of the flower — the stamens and pistil — so that the bee's body will usually come into contact with these structures as it slurps up the nectar or stuffs pollen into the little "baskets" on its rear legs. However, in doing so, it also gets the dusty pollen all over its body. When it flies to the next flower, some of the pollen from its body hairs may rub off and onto the pistil of the flower. Presto, the new flower is pollinated. Bees have evolved to gather pollen as a high-protein "prey," rather than hunting other invertebrates for protein like most wasps do. The bees use the pollen as food for their larvae in the hive. Most of the nectar, on the other hand, goes toward the production of honey, which is stored for consumption during the winter and for fuelling the bee's own energy needs.

In recent years, bee populations have been declining — in some areas quite drastically. Honey bees (*Apis mellifera*) have gotten most of the press. In Ontario, beekeepers have lost 30 to 40 percent of their bees every spring since 2006.[4] The main culprit appears to be the varroa mite, a crab-like parasite the size of a pen dot. According to Peter Kevan of Guelph University, Ontario is also losing wild bees. Two native bumblebee species that used to be the most common bees in Ontario are in a massive decline and may already be extirpated in some areas.[5] The rusty-patched bumblebee (*Bombus affinis*) has seen especially drastic declines and has now been officially designated as endangered in Ontario. This species is an important pollinator for a large variety of wildflowers. The causes behind the collapse of native bumblebee populations are largely unknown but may be related to pesticide use, habitat loss, climate change, and disease.

Bee decline has very serious implications for agriculture, since many of our key fruits and vegetables are pollinated by bees. In fact, about one-third of the human diet comes from insect-pollinated plants — mostly thanks to the work of the honey bee. Gardeners can help bees by creating pollinator-friendly gardens by growing native plants and eliminating the use of pesticides.

PLANTS AND FUNGI

APRIL HIGHLIGHTS

All Month Long
- A pastel wash of swelling buds and emerging tree flowers spreads over the landscape, giving distant trees a soft, hazy appearance. In towns and cities, the dense clusters of silver maple flowers festoon twigs with tinges of red, yellow and green.

- Scarlet fairy cup fungus (*Sarcoscypha austriaca*) fruits on twigs or small branches. It is one of the earliest of spring fungi and often hidden under fallen leaves. The cups are up to four centimetres across and a brilliant red inside.

Early April
- The yellow, dandelion-like flowers you now see growing in roadside ditches are a non-native species known as coltsfoot. Initially, coltsfoot sends up just flowers; only later in the spring will the hoof-shaped leaves appear. By month's end, the flowers are replaced by white, fluffy seed heads which also resemble those of dandelions.

- The male catkins on speckled alder bushes start to swell into long, caterpillar-like, yellow and purple garlands that release golden puffs of pollen when jostled. The female flowers are nestled in small, erect catkins that become cone-like in appearance when the seeds are ripe. Look for alders along shorelines and on the margins of wetlands.

Mid-April

- One of the very first trees to come into leaf in the spring is the weeping willow. It is often the last tree to lose its leaves in the fall, as well. Although they are attractive, non-native species such as this tend to take away from our "sense of place."

- Hepatica are usually the first woodland wildflowers to bloom in the spring. The flowers can be pink, white, or mauve. Look for them on south-facing forest hillsides, sometimes right at the base of a large tree.

- Dark brown false morel (*Gyromitra esculenta*) mushrooms begin to appear in woodlands. False morels have a wrinkled, brain-like appearance and are poisonous.

Late April

- By month's end, most of the wind-pollinated trees are in full bloom. Alder, birch, and elm pollen is especially abundant. (See page 99)

- On the Shield, watch for the fragrant, white blossoms of trailing arbutus. This plant tends to grow in exposed sandy areas, often near pine trees.

Bloodroot leaves clasp the stem when the flowers first appear. The root contains blood-red juice.
Terry Carpenter

"First Bloom" Calendar for Selected Trees, Shrubs, and Herbaceous Plants

Early April: Coltsfoot, red maple, speckled alder

Mid-April: American elm, round-lobed hepatica

Late April: Manitoba maple, leatherwood, bloodroot, blue cohosh, trailing arbutus, Dutchman's breeches, marsh marigold, violets, dandelions, peduncled sedge, distant sedge, plantain-leaved sedge, mountain rice grass

Sex in the Trees

Almost without our knowing it, the flower buds of maples, poplars, aspens, willows, alders, birches, and elms are gradually transforming the April landscape. Although these blossoms are small and somewhat inconspicuous, they give the crowns of early-flowering trees a hazy appearance that is yet another sign of spring's progress. Leaves and flowers emerge based on how much heat the tree has accumulated above some minimum developmental threshold. Botanists measure the heat in growing degree days or GDD, namely the number of degrees that the average daily temperature is above a baseline value (usually 10°C). The maximum and minimum temperatures for the day are added together and divided by two. The base temperature is then subtracted from the mean temperature to give a daily GDD. A day with a high of 23°C and a low of 12°C would contribute 7.5 growing degree days. For example, sugar maple requires anywhere from 30 to 55 GDDs before flowering will occur.

Because insect activity is unpredictable during the cool days of April, most early-flowering species depend primarily on the wind for pollination. But, since wind may blow from the wrong direction or not at all, trees that use this mechanism have hedged their bets by producing astronomical quantities of pollen. Take the birch tree, for example. Each of the many-flowered catkins produces about five million pollen grains. The tree may have thousands of these catkins. Wind-pollinated species also bloom before the leaves emerge because, if they didn't, the leaves would shield the flowers from the wind and make it more difficult for the pollen to reach the pistils. Wind-pollinated trees often have male flowers that hang like tassels in order to take advantage of the slightest puff of wind.

One of the first trees off the mark is the silver maple, whose flowers usually appear in late March. Its fat, showy clusters of flower buds are red, but the flowers themselves are greenish-yellow. Both male and female flowers appear on the same tree. About a week later, red maples blossom and brighten the landscape with crimson red flowers. Packed in tight clusters, male and female flowers usually appear on different branches of the same tree. The third maple to flower in April is the Manitoba maple, a somewhat aberrant member of the clan. Not only does it have ash-like, compound leaves, but the pollen flowers and seed flowers appear on completely separate trees. The male flowers are at the end of long, slender stalks. This design probably facilitates pollination by the wind. Most maples, however, are both wind and insect-pollinated.

Elm flowers, looking like so many brown raindrops hanging from the branches, also add new colour to early spring's palette. The small, wind-pollinated flowers are clustered in loose tassels. The seeds will mature and fall away even before the leaves reach their full size.

"LEAF-OUT" CALENDAR FOR SELECTED TREES, SHRUBS, AND HERBACEOUS PLANTS

For each of the spring and summer months, the "leaf-out" calendar gives the approximate time when the better-known species come into leaf. As with flowering, the year-to-year variation in weather may accelerate or slow down the process of leafing out. Microclimate can also have an effect. For example, cities tend to be warmer than outlying areas, so city trees are usually in leaf first.

Late April: Currants, red elderberry

WEATHER

APRIL HIGHLIGHTS

All Month Long

- We begin the month with about 12 ¾ hours of daylight and end with a little over 14.

- The smell of early spring is in the air. It is a blend of the earthy smell of soil, leaf mould, rotting twigs, earthworms, and the first spicy odours of balsam poplar buds.

- Despite the saying "April showers bring May flowers," this is not a month of heavy precipitation in central and eastern Ontario. In fact, in all areas except Kingston, April precipitation is well below the monthly average for the year.

- Vernal ponds — small, temporary bodies of spring meltwater and rain, located in depressions of the forest floor or adjacent fields — create crucially important breeding habitat for frogs, toads, and salamanders. Because they usually dry up by midsummer, these ponds are an aquatic habitat that is free of predation from fish that would otherwise eat the amphibians' eggs and young.

Early April

- Snowstorms and flooding are often part of the weather picture early in the month.

Late April

- Lake Simcoe, Lake Couchiching, and the Kawartha Lakes are usually ice-free by about April 20. The process by which the ice melts and "goes out" of our lakes is intriguing. (See below)

- April 30 is the average date of the last spring frost in Kingston.

APRIL WEATHER AVERAGES, 1971–2000[6]

City or Town	Daily Max. (°C)	Daily Min. (°C)	Rainfall (mm)	Snowfall (cm)	Precipitation (mm)
Owen Sound	9.8	0.8	61.6	8.7	70.2
Huntsville	10.3	-1.1	54.7	8.5	63.2
Barrie	10.6	0	57.8	5	62.9
Haliburton	10.3	-1.5	56.6	13.7	70.2
Peterborough	11.3	0.8	59.6	6.9	66.1
Kingston	10	0.8	74.8	9.2	84.9
Ottawa	10.9	1.1	59.7	7.5	67.7

ICE-OUT

On our lakes, winter is slow to loosen its grip. The retreat of the ice occurs in fits and starts as spring warmth grows but then falters. But eventually spring wins out, and by late April most central and eastern Ontario lakes are open. The process, however, is more complicated than you might think. From below, the ice is being melted away by the relatively warmer meltwater flowing into the lake along the shoreline. From above, the ice is eroded by the warmth of the spring sun. When the snow cover on the surface melts, the deterioration of the ice picks up speed. Solar radiation penetrates into the ice, creating tiny cracks. Meltwater filters down through these cracks, creating vertical columns. Ice that has been deteriorated in this manner is known as "candled ice" because it looks remarkably similar to a layer of vertically packed candles. With so much water present in the ice, the ice cover appears quite dark in colour. Then, quite suddenly, within 24 hours, the ice is gone, having essentially collapsed and any remaining solid ice melting away in the slightly warmer lake water. This fascinating process is also accompanied by its own music. Listen especially for the crystal tinkling of candled ice and sometimes even the noise of waves rolling under the ice surface.

APPROXIMATE APRIL SUNRISE AND SUNSET TIMES (DST)[7]

(Note: Twilight starts about 30 minutes before sunrise and continues about 30 minutes after sunset.)

Location	Date	Sunrise (a.m.)	Sunset (p.m.)
Grey-Bruce	Apr. 1	7:06	7:50
	Apr. 15	6:41	8:07
	Apr. 30	6:17	8:26
Muskoka/Haliburton	Apr. 1	7:01	7:45
	Apr. 15	6:36	8:02
	Apr. 30	6:12	8:20
Kawartha Lakes	Apr. 1	6:56	7:39
	Apr. 15	6:31	7:56
	Apr. 30	6:07	8:15
Kingston/Ottawa	Apr. 1	6:49	7:32
	Apr. 15	6:24	7:49
	Apr. 30	6:00	8:07

NIGHT SKY

APRIL HIGHLIGHTS

All Month Long

- Major constellations and stars visible (April 15 — 10:00 p.m. DST)
  Northwest: Taurus (with *Aldebaran*) in northwest; Cassiopeia is low in the sky; Pleiades are to its left.
  Northeast: Big Dipper upside down fairly high; Little Dipper (with *Polaris*) below it.
  Southeast: Leo (with *Regulus*) rules over the southern sky, Bootes (with *Arcturus*) and Virgo (with *Spica*) are almost due east.
  Southwest: Orion (with *Betelgeuse* and *Rigel*) is low in the west; Taurus (with *Aldebaran*) high to its right; Canis Major (with *Sirius*) low to its left; Canis Minor (with *Procyon*) due south; Auriga (with *Capella*) high above Orion; Gemini (with *Castor* and *Pollux*) to left of Auriga.

- Easter is celebrated on the Sunday after the first full moon following the spring equinox. The Christian Easter was preceded by an old pagan festival celebrating fertility and new growth. The date of Easter reminds us of how closely human events were often tied to the phases of the moon. Knowing the different phases adds one more dimension to appreciating the rhythms of the natural world. (See page 103)

- The Algonquian name for the full moon of April is the Pink Moon.

- Early spring is the best time of the year to see a lunar halo. Look for it on evenings when the temperature is a few degrees above freezing. The halo appears like a silvery ring around the moon and is caused by the prism-like nature of high-altitude ice crystals.

Late April
- The Lyrids meteor shower can be seen in the northeast from April 20–22.

LEARNING THE PHASES OF THE MOON

Many of us are unfamiliar with the comings and goings of the moon. It is just "somewhere up there," its movements largely ignored and its amazing surface only rarely examined closely. Like the sun, the moon rises in the east and sets in the west. It follows roughly the same path through the sky as the sun and is actually visible during the day as much as during the night. But, unlike the sun, the moon rises each day an average of 50 minutes later than the day before. It takes the moon about 29 ½ days to go through the full cycle of eight distinct phases.

1. **New moon:** The new moon rises and sets with the sun and stays close to it during the day. The sun shines on the far side of the moon during this phase, making the moon invisible.

2. **Waxing crescent:** The moon rises and sets shortly after the sun and can be quite striking in the evening twilight, low in the west. Earthshine (sunlight reflected off the Earth onto the moon and back again) dimly illuminates the moon's surface to the left of the crescent. Looking like the rounded part of a *D*, the waxing crescent is "developing" toward the full moon. This phase is poetically described as "the old moon in the new moon's arms."

3. **First quarter:** This is the familiar "half moon." It is called a quarter moon, however, because it has completed one quarter of its cycle. A quarter moon is in the sky about half the day and half the night. This is the best moon phase for looking at the moon's surface through binoculars.

4. **Waxing gibbous:** The word *gibbous* means "like a hump." The somewhat football-shaped waxing gibbous moon rises late in the day and shines most of the night.

5. **Full moon:** The beautiful full moon rises at sunset and sets at sunrise. It often appears like a huge orange ball as it climbs above the eastern horizon in the evening.

6. **Waning gibbous:** The waning gibbous moon rises after sunset and starts to take on the shape of a *C*. The moon is "crumbling" away.

7. **Last quarter:** This phase rises in the middle of the night and sets at midday.

8. **Waning crescent:** What is left of the crumbling moon rises and sets just before the sun and stays in the sky most of the day. It can be exquisitely beautiful at dawn.

May

The Promise of Spring Fulfilled

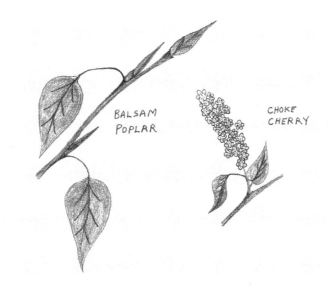

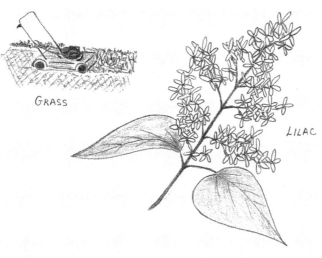

The May air is full of signature smells, such as balsam poplar, chokecherry, fresh-cut grass, and lilac.

Kim Caldwell

The world's favorite season is the spring.
All things seem possible in May.

— Edwin Way Teale

ALL OF THE PROMISES THAT NATURE HAS BEEN MAKING SINCE THE WINTER SOLSTICE are fulfilled in May. The birds of spring arrive en masse, leaves and flowers burst out all around us, and any hint of winter is soon lost in the warmth and the sunshine. Although it seems foolish to talk of a favourite month, for anyone who takes pleasure in watching the seasons unfold, May has no equal.

Not that all is idyllic. May advances in fits and starts depending on the vagaries of the weather. Cold weather is no stranger to the month, nor are days of midsummer heat. But when the warm weather does arrive, it triggers change at a dizzying pace. This can be frustrating for anyone who is attempting to observe everything new that is happening.

May begins with tree branches bare to the sky and ends with the freshness of new leaf as a green veil is drawn down upon our forests and fencerows. But as the trees leaf out, the spectrum of pastel greens, whites, browns, and reds offers a colour spectacle equal to that of fall — at least to those who take the time to appreciate the subtleties, nuances, and changes occurring from one day to the next. Woodlots are carpeted with rafts of white trilliums, fields glow with dandelion gold, and lilacs bow heavy with blossom. Although lilac may be the sweetest fragrance of the month, balsam poplar is the most pervasive. And, depending on where you live, there are also accents of cherry blossoms and freshly mown grass.

Southerly winds this month push avian migrants north to devour the billions of insects feasting on the rapidly developing green canopy. Many of these birds make brief appearances in our backyards as they make their way to nesting grounds farther north. The serene piping of White-throated Sparrows and the exuberant song of the Ruby-crowned Kinglet often provide the background music for several days as we toil in our gardens, full of expectations for the new season. The arrival of the birds of spring is no less than a reaffirmation of life.

BIRDS

MAY HIGHLIGHTS

All Month Long

- Spring migration is at its busiest this month with the arrival of the long-distance migrants from the neotropics. May is synonymous with birding at its best. (See page 108)

- With many species nesting, baby birds are inevitably found and believed to have been abandoned. Rarely is this the case. The cardinal rule is to leave them alone!

- Migrating Rose-breasted Grosbeaks, and sometimes even Indigo Buntings, occasionally drop in at sunflower feeders.

- With many species migrating through and others already nesting, try to keep your cat indoors — it is estimated that hundreds of millions of birds, small mammals, reptiles, and amphibians are killed by free-roaming cats in North America each year.

- Windows, too, take a huge toll on migrating birds. If you have a window that is particularly dangerous for birds, try drawing the blinds during the day, placing decals on the windowpane, or hanging strands of ribbon outside the window in order to reduce collisions. The migration months of May and September are the most crucial for protecting birds from window collisions.

- It is very common this month to see one or two Common Grackles chasing an American Crow, often over a considerable distance. Crows are nest robbers and grackles are one of many species who fall victim to these intelligent birds.

- Every evening during the twilight period that follows sunset, American Robins fill suburban neighbourhoods with their melodic up-and-down-the-scale song. An even more intense robin chorus can be heard in the morning, starting well before sunrise.

Early May

- Loons, either alone or in pairs, are often seen flying due north in the early morning. Even if you don't see the bird, you may hear its yodeling call, which is often given on the wing.

- Knowing the meaning of "loon language" adds a great deal to the enjoyment of these magnificent birds. (See page 109)

- Ruby-throated Hummingbirds return from Central America and make a beeline to feeders. This species is actually more common now than 20 years ago. This is no doubt partly due to the widespread popularity of hummingbird feeders and flower gardening, as well as the opening of forest for cottages. The highest abundance of Ruby-throated Hummingbirds in Ontario is along the southern edge of the Canadian Shield, which encompasses most of cottage country.[1]

- Chimney Swifts return from their wintering grounds in South America. Watch for a bird with a flickering, almost bat-like flight that is usually accompanied by a sharp, chippering call. Swifts roost and nest in chimneys. However, partly because old brick chimneys are disappearing, the Chimney Swift is now designated as a threatened species in Ontario.

Mid-May

- Songbird migration is at its peak. Migrating warblers, vireos, thrushes, tanagers, orioles, and flycatchers are most abundant between May 10 and May 25 — exactly at the right time to take full advantage of the legions of insects that are emerging.

- Blue Jay numbers increase as small flocks of migrants that spent the winter south of the border return to central and eastern Ontario.

Late May

- Common Nighthawks return from South America. The nasal *peent* of nighthawks calling on a summer evening was, until fairly recently, an integral part of the soundscape of the towns and cities of Ontario. Along with other species that feed on the wing such as swallows and swifts, nighthawks are suffering a major decline.[2]

- Most neotropical migrants will have now built their nests and laid their eggs.

- Many non-breeding Giant Canada Geese, as well as failed nesters, migrate north to the coasts of James and Hudson Bay to moult. These northward-bound flocks can be a common sight from late May through mid-June.

- Look-alike Alder and Willow Flycatchers are generally the last birds of spring migration to arrive back in central and eastern Ontario. They can only be reliably told apart by voice.

THE NEOTROPICAL MIGRANTS ARRIVE AT LAST!

With May comes the biggest push of spring migration with nearly all of the long-distance migrants from Central and South America — the neotropics — arriving. It is possible to see more species at the height of migration in May than at any other time of year. Keeping an eye on the weather forecast can improve your birding success. Unlike fall migration, north winds and cool weather in the spring stall migration. Therefore, birds often wait in large numbers for warmer air and southerly winds to "push" them along. As a general rule, however, neotropical migrants arrive on more predictable dates each year than those birds that winter in the continental United States.

An elegant synchronicity of events occurs this month. As the green canopy of leaves develops overhead, countless caterpillars emerge to feast on the verdant bounty laid out before them. And, right on cue, hundreds of millions of birds pour into central and eastern Ontario to regale themselves of this insect banquet. While some species will remain to nest, others pass through quickly as they continue northward.

Most bird species migrate at night when there is less danger from predators. Migrants generally use the daylight hours for feeding and resting. If conditions are favourable, such as with the passage of a northward advancing warm front, birds will start migrating about one hour after sunset. It is quite common at this time to hear their contact calls as they fly overhead. Changing weather conditions during the night can also cause groundings of birds. When a northward moving warm front collides with a cold front, the warm air — and the birds in it — rises over the cold.

The air cools, rain develops, and the birds are forced to land. This means that rainy mornings in May can produce superb birding.

But, why would a neotropical migrant such as a Scarlet Tanager risk a dangerous 6,000-kilometre journey to fly from Panama all the way to Ontario? In a nutshell, it is because the birds are able to raise more young. Protein-rich insects are abundant during the Canadian spring and summer, there is a much larger geographical area available, and the long days allow birds to feed their young for up to six hours longer than if they had stayed in the tropics. By flying north in the spring, they also free themselves from competition for food from tropical resident birds.

You don't have to go far afield to see the birds of May. As long as there is sufficient cover, even city backyards can have their own coterie of migrants. Habitat edges such as wooded roadsides are especially worth checking. Get out early, preferably before 9:00 a.m. By visiting different habitat types, an experienced birder can usually record 80 or more different bird species over the course of a morning in mid-May, including at least 15 kinds of warblers, three vireos, four flycatchers, and four thrushes.

The male Scarlet Tanager with its red body and black wings is one of the most striking neotropical migrants.
Karl Egressy

Loon Language

The Common Loon is probably the best-known bird of cottage country. Many cottagers are very possessive and protective of "their" loons and are understandably concerned about the birds' welfare. Knowing more about the loon's mysterious visual displays and soulful calls allows us to appreciate the birds even more.

Loons have at least four distinct calls. The "wail" is a wolf-like call lasting about two seconds. It is given by both sexes and is used to summon a mate or offspring. The "tremolo" is a vibrating, laugh-like call that lasts only about one second and seems to indicate alarm. It is given when either sex is disturbed or senses danger. The "yodel" is a long and complex call that starts with a wail and then changes into a series of yodel-like undulations. It is territorial in nature and is given by the male. This call is most common in spring and is usually heard between dusk and dawn. Finally, both sexes will also use "hoots," which are short, soft contact calls often given when the birds congregate in flocks.

Loons are also famous for their unique visual displays. "Bill-dipping" often occurs when two birds meet each other, especially in a flock, and may serve to reduce aggressive tendencies. When territorial skirmishes occur, loons will sometimes raise their body upright out of the water and tread with their feet. Males may also hold their wings out to the side and give the yodel call. Another visual display used by males during territorial conflicts involves extending the head and neck on the water. Once again the yodel call is often produced at this time.

A Common Loon performing an aggressive upright display during a territorial skirmish.
Karl Egressy

MAMMALS

MAY HIGHLIGHTS

All Month Long

- A large variety of mammals give birth this month. These include beavers, northern and southern flying squirrels, eastern wolves, groundhogs, striped skunks, white-tailed deer, and moose.

- Moles are very active. Watch for ridges and mounds of soil on lawns. The soil is pushed up to the surface as the animals tunnel through the dirt in search of earthworms and other invertebrates.

- May and early June provide an excellent opportunity to observe red fox kits playing outside the den, often in the company of their mother.

- If you have never seen a moose, now is the time to go up to Algonquin Park. They are fairly common along the side of Highway 60 in May and June, attracted by puddles of salty snowmelt from winter road maintenance operations.

- Before giving birth to their new fawns, female white-tailed deer drive their male fawns from the previous year ago out of the immediate area. This is done in order to minimize interbreeding.

- The buck white-tails' antler growth accelerates dramatically as a result of the increased daylight. Bucks tend to be somewhat secretive during spring and early summer, spending their time with other males and allowing the does a wide berth.

- If there was a bumper seed crop on the trees (e.g., white spruce, sugar maple) the previous fall, the bounty of food will have allowed much more mice reproduction than usual, even during the winter. In early spring, however, as the amount of available food diminishes, mice look for other sources of nourishment and invasions of cottages and other buildings occur. This can explain why mice numbers in homes and cottages are abnormally high some springs.

Mid-May

- Moose give birth to one or two calves, almost always during the second half of May. An island is often the nursery of choice.

- Migratory bats such as the red and the hoary return to central and eastern Ontario from overwintering sites in the southern United States.

Late May

- Deer fawns are usually born in late May or early June.

Moose are most often seen in low-lying wet areas such as bogs, ponds, and roadside ditches, especially in the early morning.

Terry Carpenter

AMPHIBIANS AND REPTILES

MAY HIGHLIGHTS

All Month Long

- Nearly all of our frog and toad species can usually be heard calling at some time this month.

- Northern water snakes mate in May by entwining themselves around each other, often on low branches near the water's edge or among rocks and boulders.

- Wetlands, roadside ditches, and even backyard swimming pool covers are often teeming with tadpoles at some point this month. Species such as spring peepers will turn into fully formed frogs in about 12 weeks. Bullfrogs, however, remain in the tadpole stage for up to three years.

- Eastern massasauga rattlesnakes usually mate in May or June. However, mating sometimes occurs as late as September.

Early May

- Calling both day and night in extremely long, fluid trills, the American toad provides one of the most characteristic sounds of early May. (See below)

- The sight of Midland painted turtles (and in some areas Blanding's turtles) basking in the sunshine is a common sight on logs and hummocks in local wetlands. Because turtles are ecothermic (cold-blooded), they can raise their body temperature only by absorbing heat from their surroundings. A high body temperature is necessary in order to digest food and to be able to hunt effectively. By exposing their entire body — including spread-out toes — to the sun, they can achieve a body temperature eight to ten degrees higher than the surrounding air.

Mid-May

- Sounding remarkably like birds, gray tree frogs deliver their song in two-second bursts, a sound I like to compare to a gentle machine gun. This species sometimes calls during the day but is most vocal on warm, damp evenings. As long as there are lots of deciduous trees and backyard ponds, tree frogs can even be heard in suburban areas.

- The banjo twangs of green frogs, the rapid, hollow tapping of mink frogs, and the deep bass notes of bullfrogs also join the amphibian symphony about this time.

Late May

- Almost miraculously, baby Midland painted turtles emerge from the ground nests, where, like many frog species, they have overwintered as frozen blocks of "turtle ice." (See November Amphibians)

TRILLING TOADS

The call of the American toad is without a doubt one of the most pleasant and typical sounds of May. Like some frog species, male toads defend territories and advertize their presence to females by singing. Their long, high-pitched musical trill can last up to 30 seconds and can be heard both day and night. When more than one male is singing at a time, each individual will usually sing at a slightly different pitch. It is also possible to watch toads as they gather in marshes and in pools of temporary, shallow meltwater for a veritable mating frenzy. The water that collects on backyard pool covers can be a breeding location, as well. Males are easy to identify because they are half the size of females, and their inflated vocal sacs stand out prominently when they call. The males will literally throw themselves on everything that moves, including the observer's rubber boots! Once a female is found, the male will hold on to her back for dear life until she finally lays her eggs, which he will then fertilize. Unlike frogs, toad eggs are laid in ribbon-like strings instead of big clumps.

FISH

MAY HIGHLIGHTS

All Month Long

- Many of our non-game species such as minnows, sticklebacks, and darters spawn in the late spring and early summer. (See below)

- In the spring, Lake Trout frequent shallow water and do not retreat into deeper water until water temperatures warm. Anglers, therefore, fish for Lake Trout in shallow water at this time of year.

Mid-May

- The fishing season for Walleye in central and eastern Ontario opens on the second or third Saturday of May, depending on the location.

Late May

- Members of the sunfish family, such as Smallmouth Bass and Pumpkinseeds, begin to spawn, depending on the water temperature.

- Fly fishing for Brook Trout is at its best. Because the water is still quite cool, the fish are often near shore and can sometimes be observed jumping for flying insects.

GETTING TO KNOW OUR NON-GAME SPECIES

In many ways, watching fish can be just as much fun as catching them. Either from shore with the help of Polaroid sunglasses or, even better, from under the water with a mask and snorkel, fish-watching can indeed be an absorbing pastime. In addition to the many well-known game species, central and eastern Ontario boasts a number of fascinating non-game fishes to observe. A perfect example is the beautiful Iowa Darter. Unlike other fish, darters spend most of their time on the lake or stream bottom, occasionally darting from one spot to another. Spring males are so vividly coloured that you might think you are looking at an escapee from a tropical fish aquarium. In May, they show blue or green bars between the brown stripes on the sides, are yellow underneath, and have reddish lower fins. To top it off, the first dorsal fin is banded in blue and red. The Iowa Darter is an ideal species for fish-watching because it spawns so close to shore. It measures about six centimetres in length.

Most species of the minnow family are also spawning this month. The Creek Chub, a popular bait minnow, spawns in clear streams when the water temperature reaches 16° to 21°C. The eggs are deposited in a mound-like nest of stones constructed by the male. It is possible to observe the male guarding the eggs from predators. This attractive minnow has a coppery appearance and a very dark lateral stripe. Spawning males actually develop large, sharp tubercles on the head. Another common minnow, the Fallfish, also constructs a mound of stones in which it nests. The male actually carries the stones in its mouth. These two species are very similar in appearance but can be told apart by the presence of a black spot at the base of the dorsal fin in the Creek Chub. Fallfish sometimes reach lengths of up to 50 centimetres. Creek Chub and Fallfish play the role of top predator in streams where trout are absent.

A species to watch for in quiet, vegetated ponds is the Brook Stickleback. In late spring or early summer, the male actually constructs a nest of small sticks and bits of vegetation. He then induces one or more females to lay eggs in the nest, which he guards aggressively. *The Royal Ontario Museum Guide to Freshwater Fishes of Ontario* (See Bibliography) is an excellent resource to fish identification and ecology.

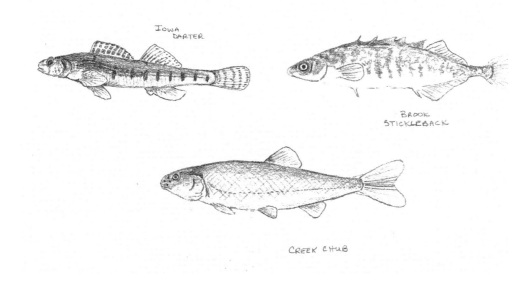

The Iowa Darter, Creek Chub, and Brook Stickleback are three of Ontario's most fascinating non-game fish species.

Kim Caldwell

INSECTS AND OTHER INVERTEBRATES

MAY HIGHLIGHTS

All Month Long

- Damselflies and dragonflies change from forbidding aquatic nymphs to gracious adult flying machines, ready to prey on other insects. Watch for early damselflies such as the boreal bluet and eastern forktail, as well as dragonflies like the American emerald, Hudsonian whiteface, beaverpond baskettail, four-spotted skimmer, and the common green darner.

- This is the flight season for the spring ephemeral butterflies such as the Olympia marble, Chryxus Arctic, the rare West Virginia white, and several of the elfins. Elfins are small brownish butterflies that fly up from gravel roads and woodland trails. Start looking for all of these species early in the month, especially if the weather is warm.

- American lady, painted lady, and red admiral butterflies arrive in Ontario from the United States. Unable to overwinter here, their presence in Ontario depends on American populations expanding their ranges in the spring and summer and spilling over into southern Canada. Their abundance varies a great deal from one year to the next.

- The spring azure butterfly is usually fairly common by the second week of May. Unlike mourning cloaks, which overwinter as adults, the azure emerges from its chrysalis in spring — one of the very first butterflies to do so. This beautiful insect is blue above and white below.

- When the water warms sufficiently, crayfish moult and then mate. In many areas of central and eastern Ontario, the native crayfish are disappearing as a result of competition from rusty crayfish, an aggressive alien species inadvertently introduced from the United States, possibly through bait bucket dumping. It is named for the reddish-brown patches on its carapace. Ontario is home to ten species of crayfish.

- On warm, breezy days from spring through fall, watch for strands of gossamer — fine filaments of spider silk — caught up in tree branches and bushes. Closer examination will often reveal a tiny spider attached to the silk. In a phenomenon known as ballooning, a newborn spider will climb up to the tip of a twig and release several silk threads from its abdomen into the air. The threads are caught by the breeze and carry the spiderling away — usually only a few metres, but sometimes thousands of kilometres. In this way, spiderlings can establish their own territory and avoid interbreeding.

Early May

- Black flies are usually at their worst about now. If it helps, just think of them as part of the price we pay for wildflowers. (See below)

Mid-May

- At about the same time as the wild cherries bloom, mosquitoes become quite notice-able and annoying. Mosquitoes feed on the nectar of cherry blossoms and help to cross-pollinate the trees. (See page 118)

- The spring field cricket serves up the first insect music of the year.

Late May

- Adult June beetles, also known as June bugs (genus *Phyllophaga*), emerge from the soil and crash full-speed into screen doors and windows at night. The larvae of this familiar insect are the much-maligned C-shaped white grubs that can damage the roots of grass and shrubs. Skunks and even crows will pull back the turf in search of a meal of grubs.

- Tent caterpillar infestations sometimes reach epic proportions in late May. (See page 119)

- Gypsy moth caterpillars, a non-native species, do considerable damage to our forests some years by eating the leaves of hardwoods such as oaks and poplars and even some conifers. It is estimated that each gypsy moth caterpillar will consume about one square metre of foliage during its life! Infestations often last several years.

- Canadian tiger swallowtails and black swallowtails appear by month's end and are sometimes seen at puddles, where they ingest salts and other minerals in the water. Male butterflies require these chemicals in order to mate successfully. Sulphur but-terflies are often seen "puddling" as well.

- Frothy white masses of spittle are now a common sight on the stems of field plants. They are created by spittlebug nymphs, which suck juices out of the plant's stem and then excrete the excess. This creates a froth that envelopes the insect, providing it with protection and preventing it from dehydrating. Adult spittlebugs are known as froghoppers.

BLACK FLIES

In May, there is a price to be paid for the bird music and flowers that we enjoy so much. The price is called black flies. Unlike the "good and bad" years of mosquitoes, black flies are bad *every* year. There really is no escaping them. Black fly larvae require cold, clear running water in which to develop. This sort of habitat is common on the Canadian Shield, making most of central and eastern Ontario a black fly paradise. And because the adults disperse widely after emerging from the pupae, avoiding areas with streams is not always a solution.

The first onslaught of black flies develops from eggs that hatched the previous fall. The larvae grow slowly over the course of the winter and then spin cocoons in the spring in which they transform into adults. A bubble of air carries them to the surface, releasing a ready-to-fly adult when the bubble bursts. It is possible to see this "popping out" taking place. In other species of black flies, the eggs overwinter and hatch in the spring. These species do not mature into biting adults until June or later. Fortunately for us, those adults that emerge in mid to late summer never reach the huge numbers of the spring black flies.

The majority of black fly species produce only one generation of adults in a year. A given female, however, is able to produce eggs two or three times during her short lifetime. In most species, a blood meal is necessary in order to provide the required nutrients for egg development. It often takes two or three blood meals for the female to fill her gut, each meal taking three to five minutes of uninterrupted eating! Luckily, not all black fly species feed on humans; many prey only on birds or other mammals. Black flies are strongly influenced by colour and find dark hues most attractive. Light-coloured clothing is therefore the best choice when heading into black fly country. It might also be of some consolation to know that both male and female black flies feed heavily on nectar to obtain the energy necessary for flying and mating. In the process, they inadvertently pollinate many species of Shield plants — most likely including blueberries![3]

MOSQUITOES

When the cherries begin to blossom in mid-May, the appearance of the first mosquitoes is not far behind. Both males and females feed heavily on the nectar of cherry flowers. Like the black fly, the female mosquito also requires a blood meal for her eggs to develop properly. Although there are 59 species of mosquitoes in Ontario, many do not bite humans and most produce only one generation of adults a year. Most of those found in central and eastern Ontario belong to the genus *Aedes*. The larvae hatch from overwintering eggs laid in mud the previous year. In the spring, these muddy areas are flooded by melting snow and rainfall and create an ideal habitat — generally devoid of predators — for the larvae to develop. A warm, damp May assures quick growth and means the breeding pools will not dry up before the adult mosquitoes emerge. Some *Aedes* species breed only once, while others, such as *Aedes vexans*, breed continuously from June until September. *Vexans* is usually only a problem in wet summers, however, because it requires muddy areas created by rainfall to lay its eggs on.

One species, the common house mosquito (*Culex pipiens*), is often found in urban areas. It breeds in any locations where water collects and stands still. The eggs develop into adults in about a week to ten days under good conditions. The house mosquito is present from spring until fall. The species overwinters as adult mated females, which find shelter in locations where the temperature remains above freezing, such as cellars, sewers, and animal burrows. With warm spring days, these females seek a blood meal and begin the cycle again.

Common house mosquitoes are believed to be the main mosquito species responsible for transmission of West Nile virus to humans.[4] A mosquito becomes infected with the virus when it feeds on the blood of a bird that is already infected. About two weeks later, the mosquito becomes capable of passing the virus on to people and animals by biting them. This species does prefer biting birds, however, with robins being their favourite target.

Despite the bother mosquitoes cause us, we all need to remember the central role they play in ecosystems. Mosquito larvae convert decomposed organic matter into living insect protein, fat, and carbohydrates and serve as food for predatory aquatic insects and fish. Adult mosquitoes are an important food source for everything from little brown bats to Chimney Swifts, both of which are increasingly threatened.

Tent Caterpillars

As green foliage transforms the landscape, hordes of caterpillars are provided with an almost unlimited source of food. Tent caterpillars, in particular, can be very plentiful. We have two species in our area — the forest tent and the eastern tent caterpillar. Only the latter actually constructs a tent. Eastern tent caterpillars emerge in mid-May from "varnish-coated" egg masses wrapped tightly around the twig of a cherry or apple tree. They spin a small silken tent in a crotch of the same tree and enlarge the tent as they grow. The caterpillars feed outside the tent during the day, using it only for resting. Even though they may completely defoliate their host tree, new leaves grow within a matter of weeks and the tree usually recovers. When fully grown, the caterpillars leave the tree and spin a cocoon in some sheltered location. By summer they will have transformed into nondescript brown moths measuring about three centimetres across.

Forest tent caterpillars feed on a much larger variety of deciduous trees than their tent-making cousins. They can therefore defoliate large areas of forest in years when their population peaks. Like the eastern tent caterpillar, they are a species native to Ontario. Every ten years or so, an outbreak occurs and can last for several years. Even trees that have been completely stripped of their leaves, however, usually re-foliate in three to six weeks. These caterpillars are distinctive because of the key-hole-shaped markings on their back. Although forest tent caterpillars do not make tents, they do spin silken threads for pathways to and from their feeding sites on the trees. When the light is right, it is possible to see hundreds of these silk highways in the treetops. At maturity, the caterpillars spin white, silken cocoons on trees, fences, buildings, and other structures. They remain in the cocoons for about ten days. By about the first week of July, they will have metamorphosed into fuzzy yellow or buff-coloured moths.

Forest tent caterpillar numbers are naturally regulated by factors such as late spring frosts (which kill the larvae), bird predation (cuckoos love them), and parasitic and predatory insects. The most important predatory insect is the Tachinid fly, which can destroy more than 80 percent of the larvae.

SOME COMMON MAY BUTTERFLIES[5]

The butterflies listed below are those species you are most likely to encounter over the course of the month by making repeated visits to the proper habitat. Some species have very specific plant requirements. West Virginia white, for example, is only found in woodlands where toothwort — its larval plant — grows. This, and other butterfly lists in the book, were prepared in collaboration with Jerry Ball, a Peterborough-based butterfly enthusiast with a great deal of experience observing these insects throughout central and southern Ontario.

Black swallowtail	Eastern comma
Canadian tiger swallowtail	Compton tortoiseshell
Clouded sulphur	Mourning cloak
Mustard white	Milbert's tortoiseshell
West Virginia white	American lady*
Cabbage white	Red admiral*
Olympia marble	Chryxus Arctic
Hoary elfin	Northern cloudywing
Henry's elfin	Dreamy duskywing
Eastern pine elfin	Juvenal's duskywing
Brown elfin	Columbine duskywing
Spring azure	Hobomok skipper

Numbers vary greatly from year to year

PLANTS AND FUNGI

MAY HIGHLIGHTS

All Month Long

- A number of native sedges, rushes, and grasses bloom at about the same time as the leaves are coming out. (See page 124)

- People often develop rashes from exposure to poison ivy in the spring. With increased carbon dioxide in the atmosphere, poison ivy is thriving. (See page 125)

- Although not as flamboyant as the colours of fall, spring offers an equally beautiful flush of colour. From the white blossoms of serviceberry (Juneberry) and cherry trees to the lime greens and pastel reds of distant woodlands, the season paints the landscape with a gentle warmth all its own.

- Spring can be said to be "early" or "late" depending on the date that leaves and flowers emerge. These dates can vary by several weeks from one year to the next. However, the order in which the different species come into leaf and flower is always the same.

- Dandelion blossoms cover lawns and fields in a mantle of yellow and provide copious amounts of pollen and nectar to insect visitors. Insects see the flowers as shining points of ultraviolet light set against a green background, which they perceive as grey. It is fascinating to watch as bumblebees will suddenly change their flight path and head directly to the dandelions like iron shavings to a magnet.

- For most of May, trees such as ash and oak look much the same as they did all winter. They will not leaf out until the end of the month.

- Sex is literally in the air as tree pollen typically reaches its highest levels. Birch, oak, and maple are only a few of the species releasing pollen from their wind-pollinated flowers at some point this month. This means itchy eyes, stuffy noses, and lots of sneezing for people with sensitive respiratory systems. Birch pollen is nearly as irritating as ragweed. Pollen also leaves a dusty yellow film on car windows and forms a yellow border around puddles.

- Late spring frosts may destroy blooms or injure flower buds so that the bloom is deferred or never develops at all.

- Morel mushrooms fruit in May and early June. The cap is shaped something like a pine cone with a series of pits and ridges. The black morel (*Morchella elata*) appears first, followed by the yellow morel (*Morchella esculenta*). Morels are edible, but always be sure of the identification, since poisonous look-alikes do exist.

Early May

- Along roadsides and rail-trails, watch for the light brown stems of common horsetails rising from the ground in large colonies. You can see the spore cones on the tips of some of the stems.

- Early in the month, serviceberries stand out like white beacons against the slowly greening landscape. These small trees grow in clumps and are especially noticeable along roadsides when their beautiful masses of white, five-petal flowers burst forth. It is believed that white is a particularly efficient colour at reflecting sunlight from a plant's petals onto the central reproductive parts of the flower. This helps to accelerate the development of pollen and seed in the cool spring weather.

- The yellow-gold flowers of marsh marigolds, also called cowslips, brighten wetland habitats.

- Watch for fiddleheads emerging among the dead leaves of the forest floor. Actually the young, coiled leaves of the ostrich fern, fiddleheads resemble the scroll at the end of a violin. Nearly all ferns have fiddleheads, but those of the ostrich fern are the edible variety.

Mid-May

- Deciduous woodlots display a profusion of spring ephemeral wildflowers, including spring beauty, yellow trout lily, large-flowered bellwort, and, of course, trilliums. Their life cycle is controlled by the rhythms of the forest canopy. (See page 126)

- Most trees are in leaf by the middle of the month.

- Sugar maples glow in a garb of pale lemon as thousands of tassel-like flowers hang from the twigs. Within a week or so, the male flowers fall to the ground, leaving a yellow confetti on sidewalks, driveways, and roadsides. Norway maples in mid-May look like giant lime-green pompoms. Their leaves and flowers emerge simultaneously.

- Stands of bigtooth aspens become quite noticeable because of the unique silver-grey foliage of the emerging leaves.

- The soft, light green needles of the tamarack are every bit as beautiful as the smoky gold colour they become in late October.

- Common lilac is usually in bloom. Lilac leaf emergence and bloom times have long been monitored in both Canada and the United States, since they are examples of natural events whose timing can be influenced by climate change. University of Wisconsin-Milwaukee researcher Mark Schwartz has calculated that lilac leaves now emerge in the northern hemisphere about five days earlier than they did 50 years ago.[6]

- Pin cherries bloom at about the same time as lilacs. A little later in the month, choke-cherry blossoms appear. Mosquitoes are attracted to cherry blossoms, as are bees, but who designed who? (See page 126)

- Oaks are in flower. The caterpillar-like male flowers (catkins) are as long as the emerging leaves. The tiny female flowers that produce the acorns bloom at the base of the leaves.

Late May

- Wild columbine is now in bloom on rocky hillsides and along roads and trails. The flowers, a beautiful blend of red and yellow, hang in a bell-like fashion and are visited by hummingbirds.

- The spring ephemeral wildflower display draws to a close. Trees have leafed out and the forest floor is shrouded in shade.

- The cottony white material floating on the May breeze that collects on lawns and sidewalks might be mistaken for snow if it wasn't for the temperature. Instead, it consists of tiny seeds, each attached to a miniscule white "parachute." They are released by trembling aspen trees.

- Elms and silver maple begin to shower the ground with their seeds.

- Yellow lady's slipper orchids start to bloom. They line the roadsides of the Bruce Peninsula and can be as common in places as dandelions!

- Calypso (fairy slipper), probably the most beautiful of all Ontario's orchids, starts to bloom on Flowerpot Island in the Bruce Peninsula.

"FIRST BLOOM" CALENDAR FOR SELECTED TREES, SHRUBS, AND HERBACEOUS PLANTS

Early May: Leatherleaf, sweetgale, fly honeysuckle, red trillium, marsh marigold, spring beauty, yellow trout lily, early meadow rue, large-flowered bellwort, prickly gooseberry, serviceberries.

Mid-May: Pin cherry, red elderberry, common lilac, white trillium, red trillium, painted trillium, wild strawberry, pussy toes, early saxifrage, two-leaved toothwort. Flower cones appear on pine, fir, and spruce.

Late May: Striped maple, white ash, chokecherry, tartarian honeysuckle, red osier dogwood, hobblebush, showy orchis, early coral root, pink lady's slipper, yellow lady's slipper, ram's head orchid, calypso orchid, fringed polygala, Jack-in-the-pulpit, wild columbine, barren strawberry, false Solomon's seal, mayapple, common buttercup, blueberries, currants, hawthorns, some mustards.

"LEAF-OUT" CALENDAR FOR SELECTED TREES, SHRUBS, AND HERBACEOUS PLANTS

Early May: Tartarian honeysuckle, pin cherry, chokecherry, Manitoba maple, Norway maple, trembling aspen, common lilac, willows, ostrich fern.

Mid-May: Sugar maple, bigtooth aspen, tamarack.

Late May: Red oak, bur oak, American basswood, American elm, white ash, red maple.

White flowers, such as these service-berry blossoms, are typical of many plants that bloom in May.
Drew Monkman

AN OVERLOOKED TREASURE — GRASSES, SEDGES, AND RUSHES

Probably the most overlooked group of plants, even by naturalists, is the grasses, sedges, and rushes. This is unfortunate because, to many people's surprise, these plants do have flowers, the flowers are surprisingly colourful, and there is a great deal of diversity in their size and design. Several problems, however, exist for the amateur naturalist. Most plant books base identification on characteristics of the mature fruit, which are quite distinctive. However, fruit matures later in the season (June for many sedges and July or August for most grasses), making early spring identification more difficult. Once you get to know the species well, however, identification based on flower or vegetative traits becomes possible. It is also necessary to use the scientific names when dealing with these plants, since many have no common name.

A good starting point for learning about this somewhat neglected area of our flora is to memorize the following mnemonic. It will come in handy the next time you try to tell whether a given clump with grass-like leaves is actually a grass, instead of a sedge or rush. It goes like this:

> Sedges have edges,
> Rushes are round,
> Grasses are hollow,
> What have YOU found?

To tell if you are looking at a sedge, try a simple touch test. When you rotate a sedge stem between your fingers, you can feel a definite triangular shape and three distinct edges. Most rushes, on the other hand, have very round stems, while the stems of grasses are somewhat flattened and hollow inside. Grasses also "have nodes all the way to the ground" — an alternative for lines three and four of the mnemonic. Nodes are the conspicuous raised areas of the stem — think of bamboo — where leaves are attached.

A number of sedges, grasses, and rushes flower in the spring. However, the flowers are relatively inconspicuous and usually yellow-green, rust, or tan in colour. The earliest species are often in well-drained upland woods. Pennsylvania sedge (*Carex pennsylvanica*) is a common spring-flowering sedge species in more southern parts of central and eastern Ontario. It is replaced by the very similar distant sedge (*Carex lucorum*) on the Canadian Shield. Both species grow in open woodlands. Two other rather common spring-flowering sedges are plantain-leaved sedge (*Carex plantaginea*), a distinctive broad-leaved sedge, and peduncled sedge (*Carex pedunculata*), a smaller species with reddish-based flowering stems. Both species occur in a variety of woodland habitats.

Common spring-flowering grasses include woodland poa (*Poa alsodes*), a species of upland woods, and mountain rice (*Oryzopsis asperifolia*), which grows in a variety of wooded and semi-wooded situations. A less common species is sweetgrass (*Hierochloe odorata*). It can be found along roadsides and on open, often sandy ground. This is the sacred sweetgrass that many Native people of North American use for smudging (taking the smoke in one's hands and rubbing or brushing it over the body). Although most rushes flower in late summer, two wood rushes, *Luzula multiflora* and *Luzula acuminata*, are spring-flowering. Both grow in woodlands.

POISON IVY

Poison ivy also attracts considerable attention in the spring, but not because of its flowers. May seems to be the time of year when so many of us end up with its infamous rash. There are three easy ways to distinguish poison ivy from other three-leaved plants. The middle leaflet has a much longer stem than the other two, the leaflets droop downward, and at least one of the leaflets is almost always asymmetrical — the left side and right side are different. For example, one side may have three lobes and the other side none. There is great variability in leaf size and shininess and also in the sort of habitat where the plant is found. Researchers at Duke University in North Carolina have found that poison ivy proliferates as carbon dioxide concentrations in the air increase and heat up the atmosphere.[7] When high levels of CO_2 were pumped into test plots, they discovered that the plant grew 149 percent faster and became bigger. As if this was not enough, the CO_2-rich environment also produced a concentration of urushiol that was 153 percent higher than poison ivy grown in control plots. Urushiol is the resin responsible for the rash. About 70 percent of people are quite allergic to poison ivy, while the other 30 percent may eventually become susceptible after repeated exposures. A rash from poison ivy can be contracted in a number of ways. These include exposure to smoke from burning the plants, simply touching an unbroken leaf, or even petting the family dog after he has gone for a romp through the plants.

WHO DESIGNED WHO?

A number of the trees and shrubs that flower in May have harnessed bees as their mating agents by developing flowers whose shape, colour, scent, and food rewards are tailored to these insects. The bright white colour and lovely perfume of cherry flowers, for example, are like beacons, advertizing to mosquitoes, bees, and other insects from afar and inviting them to come for delicious rewards of nectar and pollen. As the bee roams the flowers' surface feeding on nectar, it becomes dusted with pollen, which it scrapes from its body into special baskets on the legs. The pollen will be fed to the larvae back at the hive. However, some pollen remains on the insect's body hairs and is inadvertently transferred to the stigmas of other flowers on other cherry trees that the bee visits. As you watch all the insect activity around a serviceberry or cherry tree, take a moment to reflect on this wondrous accomplishment of evolution. The tree, for its part, has tailored every aspect of its flowers to attract insects. In turn, the insects are totally dependent on flowers for their own existence. Their bodies and behaviours are perfectly adapted to take full advantage of everything the flower has to offer. Neither organism can exist without the other. In essence, insects designed the flowers and the flowers designed the insects.

LIVING BY THE RHYTHMS OF THE FOREST CANOPY

Most people who enjoy nature would agree that the spring ephemerals — woodland herbaceous plants with a short blooming period — are the most beautiful and finely adapted segment of the yearly floral calendar. This group includes spring beauty, mayapple, blue cohosh, yellow trout lily, Jack-in-the-pulpit, bloodroot, squirrel corn, Dutchman's breeches, violets, toothworts, hepaticas, and trilliums. They are nearly all restricted to deciduous forests and woodlots, because their life cycle is attuned to the rhythms of the forest canopy. For this group, the time available to flower and to complete seed maturation is very short. Once the forest canopy closes, the light available on the forest floor for photosynthesis falls to 1 percent of the level at the top of the canopy. In the deep shade of the inner forest, plants struggle to produce enough food even for their own needs, let alone have food left over for the process of seed maturation.

Plants must also face the challenge of attracting pollinators to carry out cross-pollination. Cool, damp weather reduces insect activity to almost zero and, should the cold continue, can seriously jeopardize the possibility of seed production. In mid-spring, there are relatively few pollinators to go around, even at the best of times. Flowers initially attract the attention of pollinators by their shape, colour, and scent. They then offer up "floral rewards" of nectar and pollen. Some species, like Jack-in-the-pulpit and red trillium, use their colour and putrid odour to deceive pollinator flies into thinking that they are pieces of rotting meat. The majority of woodland plants, however, are not overly specialized and they attract a broad array of insects. Most flower for only about two weeks but, being long-lived perennials, their seed success in any one year is less important than their success over a multi-year period.

The painted trillium has a pink ring in the middle that is thought to guide insects to a reward of nectar.

Terry Carpenter

127

In order to be ready to take full advantage of the short photosynthetic season available in a deciduous forest, spring woodland plants actually preform their flowers in miniature the year before. The flowers form in a bud at the tip of the underground rootstalk, but their cells are compact and unexpanded. Expansion into full, above-ground flowers is mostly the result of water uptake when warm spring conditions arrive. Leaves are preformed in a similar manner, but their full expansion still requires some new photosynthesis to take place. New photosynthesis is also necessary for seed maturation and to provide stored food reserves for the next year's growth. Once the forest canopy is covered in leaves, many spring ephemerals completely wither away and die back to their underground parts of roots, rhizomes, and bulbs.

WEATHER

MAY HIGHLIGHTS

All Month Long

- Our lakes receive a much-needed breath of oxygen during the so-called "spring turn-over." (See page 129)

- We gain over an hour in daylight this month. We begin with about 14 ¼ hours of daylight and end with nearly 15 ½.

- Don't put your winter parka away just yet. May can serve up anything from mid-summer heat to chilling wind, rain, and even snow.

- The air is rich with the fragrance of balsam poplar resin, a characteristic smell of spring throughout central and eastern Ontario. The scent originates from the sticky resin that oozes from the tree's buds as they open. In the city, the smell of lilac blossoms and freshly mowed grass mixes nicely with the balsam. A researcher in Australia, Dr. Nick Lavidis, has found that a chemical released by mowed grass can help people relax and put them in a better mood![8]

- There is generally a period of summer-like weather. This brings about an explosive greening of the landscape and all of the related effects on our fauna. Trying to keep up with all the "firsts" of the season can be a frustrating experience!

- The waters of the Great Lakes are slower to warm than the adjacent land. This so-called "lake effect" creates cooler temperatures along the shores of the Great Lakes than at points farther inland. This, in turn, delays leaf-out on trees. Lake effect is particularly pronounced on the Bruce Peninsula.

Mid-May

- The last spring frost occurs in many areas. The average date of the last frost is May 11 for Owen Sound, Barrie, and Ottawa and May 17 for Huntsville, Haliburton, and Peterborough. The average date for Kingston is April 30. For most of central and eastern Ontario, frost will not occur again until late September or early October.[9]

MAY WEATHER AVERAGES, 1971–2000[10]

City or Town	Daily Max. (°C)	Daily Min. (°C)	Rainfall (mm)	Snowfall (cm)	Precipitation (mm)
Owen Sound	16.6	6.1	71.8	0.2	72
Huntsville	18	5.7	79.3	0.6	79.9
Barrie	18.1	6.5	77.3	0.1	77.3
Haliburton	18.1	5.2	91.1	1.8	92.8
Peterborough	18.8	7.0	77	0.5	77.5
Kingston	16.5	7.1	74.9	0.3	75.2
Ottawa	19.1	8	80.9	0.2	81

THE SPRING TURNOVER

After the ice retreats from lakes in the spring, a critically important phenomenon occurs. The lake is allowed to "breathe" for the first time since the previous fall. Aquatic plants, however, which create oxygen through photosynthesis, do not do the job on their own; they contribute only a small amount of this essential gas. Fortunately, there is a brief period in the spring and fall when oxygen from the air above can enter the lake. At these two times of year, all of the water in the lake is at more or less the same temperature. This permits a thorough mixing of the water. On windy days, wave action on the surface creates currents which extend from the surface right to the bottom of the lake. Oxygen from the air mixes with the water and is carried throughout the lake by these currents.

But, as summer approaches, the surface water heats up considerably while the water deeper down remains at 4°C. Because warm water is lighter than cold water, the two do not easily mix and currents cannot penetrate the colder layer below. Oxygenation of this lower level therefore stops. Virtually no mixing of the two water masses will occur until the fall, when the upper level of the lake once again cools down to the same temperature as the water below.

Approximate May Sunrise and Sunset Times (DST)[11]

(Note: Twilight starts about 30 minutes before sunrise and continues about 30 minutes after sunset.)

Location	Date	Sunrise (a.m.)	Sunset (p.m.)
Grey-Bruce	May 1	6:15	8:27
	May 15	5:57	8:43
	May 31	5:43	9:00
Muskoka/Haliburton	May 1	6:11	8:21
	May 15	5:52	8:38
	May 31	5:39	8:54
Kawartha Lakes	May 1	6:05	8:16
	May 15	5:47	8:32
	May 31	5:33	8:48
Kingston/Ottawa	May 1	5:58	8:08
	May 15	5:40	8:25
	May 31	5:26	8:41

NIGHT SKY

MAY HIGHLIGHTS

All Month Long

• Major constellations and stars visible (May 15 — 10:00 p.m. DST)

Northwest: Gemini (with *Pollux* and *Castor*) in mid-sky; Cassiopeia low in north; Auriga (with *Capella*) to its left.

Northeast: Ursa Major high in sky; Ursa Minor (with *Polaris*) below it; Bootes (with *Arcturus*) high to right of Dipper; Corona Borealis just below Bootes.

Southeast: Bootes (with *Arcturus*); Virgo (with *Spica*).

Southwest: Leo (with *Regulus*); Corvus (the Crow) to lower left.

• Every spring and fall, migrant birds use the stars as an important source of directional clues.

• The Algonquian name for the full moon of May is the Flower Moon.

Early May

• The Eta Aquarids meteor shower peaks around May 5. It radiates from the Aquarius constellation low in the southeast and is most visible after midnight. The Eta Aquarids are a light shower, averaging about 10 to 20 meteors per hour at their peak.

JUNE

Endless Days and the Urgency of Life

**Bluegills spawning
— note "cuckolder"
male in background**
Kim Caldwell

Generous June is a wonderful time to be alive and know this sweet land.

— Hal Borland

IN JUNE THE WORLD IS BRAND NEW. NEVER IN THE YEAR IS THE FOLIAGE FRESHER, the kaleidoscope of greens more vivid, the smells of the natural world more alluring, and the urgency of life more palpable. Growth and procreation are the order of the day. June's long days convey a sense of unending time. Hal Borland spoke of June as "long, sweet days we bought and paid for with long, cold nights and short, bitter days at the dark turn of the year."

This is a wonderful time of year to enjoy all that our senses can perceive. The smell of the early June air, especially on a damp morning, conveys the essence of a green world. This is a time to look closely at the beauty of new leaves and to reflect on all the services that leaves render. Not only do they provide the basic foodstuff for all animal life on earth, but they cool the air through evaporation, absorb carbon dioxide, give off oxygen, and simply make us feel good as we enjoy their gift of shade.

Although June brings new opportunities for the naturalist, this is a time of relative calm after the hectic days of May. There is still much to be seen, but the pace of change has slowed, especially since spring migration has finally ended. Insects receive more attention, some for their beauty but others for their nuisance factor! Plants, too, take over the spotlight with highly-sought species such as orchids blooming throughout the month.

In June we have the sense that this time of long days and short nights will last forever. The sun arches high overhead, giving more than 15 hours of daylight. On or about the 21st of the month, we celebrate the summer solstice as the sun rises and sets farther north than on any other day of the year and thereby signals the imperceptible transition into summer.

BIRDS

JUNE HIGHLIGHTS

All Month Long

- June through early July is a critical time for loons. The birds are very vulnerable to disturbance by humans as they attempt to nest and care for their chicks. (See page 134)

- Breeding Bird Surveys are carried out to monitor the number of species and individual birds that are nesting along a given route. Volunteers drive along a predetermined 40-kilometre roadside route and make 50 three-minute stops. At each stop, the total number of bird species seen or heard is recorded. I usually record about 65 species on my route, which is located east of Peterborough between Lasswade and Havelock. The data is used to identify trends in breeding bird populations, such as the decline of grassland species. (See page 134)

- June is the time of peak nesting activity for migrants from the tropics. Most songbirds spend about two weeks incubating their eggs and another two weeks feeding their young before they leave the nest. Many species will also make attempts at re-nesting if their nest is destroyed or if predators eat the eggs or young.

- Male hummingbirds can be seen doing their pendulum courtship flight. Almost as if suspended from a string, the male flies in wide arcs above and to both sides of the female.

- Osprey eggs usually hatch during the first half of June. The eggs do not hatch all at once; instead, the first chick hatches out up to five days before the last one. The older chick dominates its younger siblings, and often eats the lion's share of the food, virtually all of which is provided by the hard-working male. Younger siblings may die if food is scarce. Ospreys continue to be common in central and eastern Ontario. In fact, their range is expanding both south and east.[1]

- Endangered Piping Plovers are nesting at Great Lakes locations such as Sauble Beach on Lake Huron.

Early June

- The last migrants pass through in the first week of June. These include Arctic-bound shorebirds such as the Semipalmated Plover, Dunlin, Least Sandpiper, and Semipalmated Sandpiper.

- In the downtown core of many towns and cities, Chimney Swifts are putting on quite a show. Pairs can be seen and heard in courtship flight as they raise their wings and glide in a *V* position.

- Birdsong is at its strongest and most diverse. Make a point of getting up and listening to the "dawn chorus," the fervent birdsong that takes place each morning before the sun comes up. This is when ownership of nesting territories is most vigorously advertized through the medium of song. The exuberance of the various songsters usually peaks at about half an hour before sunrise. Picking out the individual species at this time becomes quite a challenge but lots of fun.

Mid-June

- Many species of ducks such as Mallards begin moulting.

- This is a disastrous time for birds like Bobolinks and Northern Harriers that nest in hayfields. Most of the young are lost to mowing operations.

Late June

- With nesting duties completed, some crows and Red-winged Blackbirds are already beginning to reform flocks.

PROTECTING OUR LOONS

Loons are especially vulnerable at this time of year. The birds are incubating their eggs for most of the month and are easily scared off their nests by human disturbance. This can result in nesting failure due to cooling of the eggs or predation by gulls or raccoons. Boaters can help to protect loons by staying away from the wilder areas of lakes where the birds are nesting and by slowing down when travelling along shorelines. Boats can scare the incubating bird off the nest, and the waves can swamp the nest and drown the eggs.

The eggs hatch in late June or early July and the young almost immediately leave the nest. Once again, boat waves can cause the chicks to become separated from their parents and to fall prey to predators. Human disturbance also means that the parents are forced to spend more time keeping track of their chicks, which allows them less time to hunt for food. The chicks will not be able to catch food for themselves for at least six weeks. Despite these factors, Ontario's Common Loon populations are generally stable. There is even a small expansion of the breeding range south of the Canadian Shield.[2]

THE UPS AND DOWNS OF ONTARIO BIRD POPULATIONS[3]

First, the good news about how Ontario's birds are faring: data recorded during the Breeding Bird Atlas project (2001 to 2005) indicate significant increases in raptors (e.g., Merlin, Bald Eagle), resident species (e.g., House Finch, Northern Cardinal), and some introduced species (e.g., Trumpeter Swan, Wild Turkey). Although more forest-dwelling birds have increased than decreased in the past 20 years (e.g., Blue-headed Vireo, Pileated Woodpecker), there are still many forest species that are declining. Cerulean Warblers and Red-headed Woodpeckers are seeing an especially precipitous drop in their numbers.

Those of us who feed birds in winter may have noticed a newcomer on the scene. Red-bellied Woodpeckers, formerly restricted to the extreme southern part of the province, have dramatically expanded their range northward into many parts of central and eastern Ontario, including the Bruce Peninsula and Prince Edward County. Big birds, too, seem to be faring well, with 11 of Ontario's 12 heaviest birds showing a marked increase in the past 20 years. Among these are the Sandhill Crane and Turkey Vulture.

However, a major decline is occurring in birds that depend on grassland and open country habitat. These include birds such as Killdeer, Upland Sandpiper, Vesper

Sparrows, and Bobolinks. Grassland can be an old field no longer being farmed or even a hay meadow. The decline is partly due to the fact that woody vegetation is invading grasslands in much of eastern North America, thereby making the habitat unsuitable for these species. In addition, hayfields where some grassland species such as Bobolinks nest are often mowed during the breeding season, which destroys the nests. According to John McCracken, national program director of Bird Studies Canada, Bobolinks nesting in hayfields would see their nesting success increase from near zero percent to 80 percent or more if farmers delayed cutting their hay until the end of the first week of July.[4] Grassland birds are also threatened by changing agricultural practices such as intensification and its high inputs of fertilizer, pesticides, and mechanization. Small fields, which formerly produced a variety of crops (including pasture, hay, and small grains), have given way to large monocultures that are of little benefit to grassland species such as Bobolinks and meadowlarks. Not only is pasture plowed under, but the grassy borders of fields and roadside areas are also tilled. In addition, urban expansion and sprawl is covering former open space and agricultural land that grassland birds once used.

Cottagers often comment on the fact that there doesn't seem to be as many swallows as before. This is because aerial foragers — birds that feed on the wing by catching flying insects — are also experiencing a serious decline. In Ontario, all ten species belonging

The Eastern Meadowlark is one of many grassland and open-country species that are declining in central and eastern Ontario.
Karl Egressy

135

to this group declined between the first Atlas (1981–85) and the second (2001–05). No species increased. In fact, Common Nighthawks, Chimney Swifts, and Bank Swallows showed a greater decline than all other species of birds in the province. Whip-poor-wills, Eastern Kingbirds, and Olive-sided Flycatchers, along with all of the other swallow species, were not far behind. The most evident declines tend to be at the northern parts of species' Ontario ranges, and inland from the Great Lakes. Seven of Ontario's ten aerial foragers are also declining right across their North American breeding ranges. Since swallows are primarily birds of open habitats, natural reforestation of formerly open areas in much of central and eastern Ontario is probably a factor in their decline. Some preliminary research also indicates a correlation between diet and population drop. It is possible that the populations of some of the flying insects on which these birds feed are decreasing because of causes such as contaminants in the environment and pesticide use. "True" bugs (insect species of the order Hemiptera) and beetles seem to be particularly important to the diet of aerial foragers.

MAMMALS

JUNE HIGHLIGHTS

All Month Long

- Observing bats can be a fascinating activity, especially on warm June evenings. (See page 137)

- Beavers are very active at dawn and dusk and easy to observe. Dams are maintained throughout the year, but most new material is added during periods of high water, which usually occur in the spring.

- Female white-tailed deer have a tendency to "rule the woods" because their fawns are young and vulnerable.

- Black bears can sometimes be seen in hayfields, grazing on clover and alfalfa.

- Bears mate from early June to early August with the peak period between mid-June and mid-July. In what is probably done to show dominance and sexual readiness, bears will bite and claw the bark of trees to a height of almost two metres above the ground. They will also rub up against the tree with their haunches and in doing so leave behind a scent mark.

- Bats give birth in communal maternity roosts consisting mainly of adult females and their young. Males form their own colonies. Little brown bats will typically have one pup, while big browns usually have two. Given the precarious state of bat populations in Ontario, every effort should be made not to disturb these colonies.

- Skunks feed primarily on insects in early summer. The holes they dig in lawns are often a source of frustration for homeowners.

Early June

- Mother chipmunks force their young to leave the den to find their own territories. This dispersal is accompanied by a constant barrage of territorial *chuck-chuck* calls that can go on for days at a time.

- Baby groundhogs emerge from burrows. Watch for young raccoons, striped skunks, and red foxes, too.

OBSERVING BATS

Late spring and early summer is an excellent time of the year to observe bats feeding. Let the presence of insects be your guide. Try to find a strong, isolated white light in a rural area, preferably near a body of water. These lights often attract large numbers of insects, which in turn attract bats. Bats forage most heavily starting at dusk and continue for about an hour. They are also active again just before dawn. Little browns and big browns often forage together at the same light but are easy to tell apart because of the difference in size. Little brown bats also feed in large numbers over water just after dusk. They usually fly just above the water's surface and circle repeatedly. As darkness falls, however, it becomes increasingly difficult to actually see the animals. In this situation a bat detector (an instrument that allows you to listen to echolocation calls) is very useful. You soon become aware that there are far more bats than you expected and just listening in on the action can be most exciting.

In the relationship between bats and their prey, things are not as one-sided as it appears. Insects have developed a number of strategies to avoid becoming bat food. For example, ears that are sensitive to the bat's echolocation calls allow some moths to detect the presence of bats long before the bats detect them. Some tiger moths actually produce their own ultrasonic sounds which serve as a defence mechanism. These sounds may advertise the fact that the moths are poisonous (as larvae, many Arctiidae feed on milkweed), thereby serving as an acoustic version of the monarch butterfly's bright colours and slow, conspicuous flight. The sounds may also startle the bats, since they are not emitted until the bat is within one metre of the moth. A bat's diet, however, is not restricted to moths. They tend to be fairly opportunistic feeders and will eat whatever insects are available.

AMPHIBIANS AND REPTILES

JUNE HIGHLIGHTS

All Month Long

• Painted turtles and snapping turtles are often seen along roadsides, rail-trails, and in other sandy locations where they lay their eggs. (See page 139)

• Unfortunately, very few turtle eggs survive the 90-day incubation period. Predators usually devour the eggs within a matter of hours. Cars, too, are a major threat to turtles moving to and from nesting areas. Not surprisingly, Ontario's turtle populations are increasingly at risk. (See page 140)

• The green frog's banjo-like *poink* is a widespread sound across the region. In more northern wetlands, the "hammering" calls of mink frogs are commonly heard. These two species call both day and night.

• Milk snakes and northern ring-necked snakes lay their eggs this month and nest.

Early June

• Just as tiny spring peepers are winding their chorus down, huge bullfrogs begin to herald the approaching summer with their deep, harrumphing *jug-o-rum* calls. Bullfrogs will defend their territory vigorously, even to the point of wrestling with an intruding male.

• Five-lined skinks, Ontario's only lizard, mate in early June and are therefore more active and visible. Look for them on open, south-facing Canadian Shield rock outcrops with deep cracks. They are most easily found by carefully overturning logs and flat rocks. Be sure to carefully replace the rocks afterward and to refrain from using them to build inukshuks and cairns, which degrades skink habitat. Males in breeding condition have a bright orange chin and jaw.

Mid-June

• The gray tree frog chorus of melodious, bird-like trills usually reaches its peak about this time, providing the music of June nights. Cottagers may be familiar with tree frogs as the nocturnal visitors that gather around the porch light. Their sticky toe-disks allow them to literally walk up and down the walls and windows. A tree frog will often sit for hours without moving and then suddenly jump a metre or more, land safely, and be seen stuffing a hapless moth into its mouth. Tree frogs will continue to call sporadically all summer long.

- Most massasauga rattlesnakes in Ontario mate from mid-June to August. Males may engage in ritualized combat for access to the female. Pregnant females then move to crucially important gestation sites — micro-habitats that offer cover and easy access to a wide range of temperatures in order to incubate the young developing within their body.[5]

Late June

- Red-backed salamanders lay their eggs in rotten logs. Unlike most salamanders, they are completely terrestrial. The eggs are guarded by the adults throughout the summer. The young actually go through the larval stage in the egg and therefore hatch out as miniature adults.

- The first young spring peepers and wood frogs complete their development and emerge from ponds. They can often be found in damp, shaded locations on the forest floor.

A gray tree frog in full song, with its vocal sac clearly visible.
Joe Crowley

EGG-LAYING TIME FOR TURTLES

Starting in early June, turtles are a familiar sight along many of our roads and rail-trails. Painted turtles, followed later in the month by snapping turtles, search out nesting sites, preferably with well-drained, loose, sandy soil or fine gravel. The female scrapes out a hollow with her hind legs to a depth of about ten to 15 centimetres. Painted turtles lay five to ten white eggs, elliptical in shape and about two centimetres long. Snapping turtles lay anywhere from 15 to 30 spherical eggs that look remarkably like ping-pong balls. They are about 2.5 centimetres in diameter and vary in colour from white to pink. Although most turtles lay only one clutch of eggs, painted turtles will occasionally lay two clutches. When the female turtle has finished laying, she uses her hind legs to fill in the hole and then drags her shell over the nest to cover up any signs of her presence. Curiously enough, the eventual sex of the baby turtles depends on the temperature at which the eggs are incubated. Warmer temperatures (30°C and greater) result in all females being born, while cooler temperatures (22°C–26°C) produce only males. However, if the weather is too cool, the eggs will not be sufficiently incubated and will not hatch at all. Those eggs that are lucky enough to survive egg-eating predators such as skunks and raccoons will usually hatch in late summer.

Snapping turtles will sometimes lay their eggs a considerable distance from the nearest water. The snapper is now classified as a species at risk.

Rick Stankiewicz

TURTLES AT RISK

Sadly, seven of Ontario's nine turtle species have been classified by the Ministry of Natural Resources as Species at Risk.[6] Of these, the situation for wood and spotted turtles is so critical that they are both listed as endangered, meaning they face imminent extinction or extirpation. Both the Blanding's and eastern musk turtles are classified as threatened. Only two turtle species — the Midland painted turtle and the snapping turtle — are commonly seen, but even the snapping turtle, along with the northern map turtle, are designated as species of Special Concern. One species, the red-eared slider, is a non-native turtle that is frequently sold at pet stores. Unfortunately, disenchanted pet owners continue to release them into the wild, where they represent a threat to native species. Most released sliders also end up freezing or starving to death.

Turtle populations are in decline for a number of reasons. First of all, turtle eggs stand a very poor chance of surviving the long incubation period. Predators such as raccoons and skunks usually discover the nests within a matter of hours, dig up the eggs, and enjoy a hearty meal. They leave behind a familiar sight of crinkled white shells scattered around the nest area. Since these predators tend to flourish anywhere there is human settlement — raccoons are much more abundant than 50 years ago — very few turtle nests go undiscovered. Roadkill, too, is a very significant cause of turtle mortality, especially during the nesting season as turtles cross roads on their way to nesting sites. Killing pregnant females not only removes reproductive adults from the population, but it also means all their potential future offspring are lost as well. According to Dr. Ron Brooks, professor at the University of Guelph, even a loss of 1 to 2 percent of adults annually from the "extra" mortality of roadkill will eventually lead to the disappearance of local populations.[7]

So, what can drivers do? It's mostly a matter of slowing down and watching the road carefully when you're driving, especially when travelling near wetlands or rivers. If you see a turtle on the road, consider stopping and moving it to the shoulder in the direction it was heading. However, make sure that there is no danger from oncoming traffic before you perform the rescue. As for the snapping turtle, the safest technique is to push it along with a stout stick or lift or pull the animal, holding onto the rear of the shell.

Since 2002 the Kawartha Turtle Trauma Centre (KTTC) has been caring for injured native turtles and releasing them back into their natural habitat. Because so few turtles ever reach sexual maturity — female snapping turtles don't reproduce until the age of 18 — each adult turtle is part of an extremely important group. This is why it is so important to rehabilitate as many injured turtles as possible — especially females — and return them to the wild. Turtles can live and breed for many years and thereby perpetuate the species. If you find an injured turtle, carefully place it in a well-ventilated plastic container and contact KTTC at (705) 741-5000 for the most current list of drop-off locations.

The spotted turtle is close to disappearing from Ontario.
Joe Crowley

FISH

JUNE HIGHLIGHTS

All Month Long

- Common Carp are spawning. They can be observed throughout much of the late spring and early summer thrashing on the surface of shallow rivers, lakes, bays, and backwaters. Carp often jump right out of the water and, at times, are in water so shallow that their bodies are almost completely exposed. It's even possible to see groups of ten or more males pursuing the same female in a race to fertilize her eggs.

Early June

- Pumpkinseeds and Bluegills are spawning and display fascinating mating behaviour. (See below)

- Smallmouth, Largemouth, and Rock Bass, as well as Black Crappies and Brown Bullheads are also spawning. Male bass can be seen guarding their shallow-water nests, which are often located in the vicinity of docks. (See page 143)

- Muskellunge season opens on the first Saturday of the month in much of central and eastern Ontario.

Late June

- Bass season opens on the fourth Saturday of June in most areas.

SPAWNING SUNFISH

Central and eastern Ontario's two sunfish species, the Pumpkinseed and the Bluegill, display some fascinating breeding behaviours in the spring. Male Pumpkinseeds construct their nest in the shallow water of lakes and ponds in late May or June when water temperature reaches about 13°C. The nests are actually built in colonies ranging from just a few individuals to as many as 10 to 15. The male sweeps away gravel and other debris from the bottom with his caudal fin, almost as if he were using a whisk brush. At the same time, he holds his side fins out and pushes water forward so as to remain stationary. The male sunfish, like the male bass, is very aggressive at spawning time and will chase off intruders by charging and sometimes even nip at a bather's legs. The females remain in deeper water until the nests are completed. The male will actually swim out and "greet" an approaching female and try to drive her into his nest. If he is successful in attracting her, there is an elaborate courtship between the pair in which the two fish swim in a circular path, side by side, with their bellies touching. As the female expels the eggs, the male fertilizes them with his sperm. Not only will females often spawn in more than one nest but more than one female may also use the same nest. Although the female quickly leaves the nest after spawning, the male remains to vigorously defend the eggs and fry.

Bluegills have similar reproductive behaviour, although they breed later in the spring, when the water warms to at least 19°C. Their colonies are larger, sometimes numbering 40 or 50 nests. Female Bluegills produce an average of 12,000 eggs! However, not all Bluegill males are territorial and defend a nesting site. Some males rely on a different mating strategy. Known as satellite or "sneaker" males, they are smaller in size and will actually slip into a territorial male's nest when the female is spawning there. The "cuckolder" will then release his sperm at the same time as the larger territorial male does. Sneaker males become sexually active at an earlier age than territorial males.[8]

BASS — CARING FATHERS

Many cottagers are familiar with the sight of a large bass spawning in the shallow water at the end of the dock throughout much of June. Their aggressive nature at this time of the year is well-known, especially to bathers who may have experienced an unexpected thump on the leg. Both the smallmouth and largemouth display similar spawning behaviour. However, the smallmouth chooses rocky, gravelly sites to spawn while the largemouth prefers a mud bottom, often where water-lily roots have been exposed by the constant tail-sweeping of the male. Spawning takes place when the water temperature reaches 16°C to 18°C, usually sometime in June. The male constructs the nest at a depth of about one metre. He uses his caudal fin to "sweep" a shallow depression free of loose silt and debris. The male and female then engage in pre-spawning rituals that involve rubbing and nipping each other. Eventually, the two rest on the bottom of the nest, where the female deposits her eggs and the male releases his sperm. The female is subsequently driven off by the aggressive male. As many as three different females, however, may deposit eggs in the same nest. The male remains to jealously guard the eggs and later to protect the young fry. After they hatch, the young bass swim in a school close to the nest with the male close by. He continues to perform bodyguard duties for about two weeks.

INSECTS AND OTHER INVERTEBRATES

JUNE HIGHLIGHTS

All Month Long

- Adult mayflies — fish flies, to some people — emerge from lakes and streams and sometimes form large mating swarms, usually early in the morning or in the evening. Adult mayflies exist solely for the purpose of laying eggs. They only live for a day or so and don't even eat.

- Damselflies and dragonflies become very common and will remain so all summer. On occasion, thousands of individuals of a single species will emerge within a few days and fill the air for kilometres around.

- Aquatic invertebrates are very active and plentiful, making this a great time of year for pond studies. Try looking on the underside of rocks in streams or sweeping a net through the submerged vegetation of a pond or marsh. Many of the species you will find are in the larval stage of their development. (See page 145)

Early June

- Spectacular giant silk moths like the Cecropia take wing in June. (See page 145)

- The first monarch butterflies of the year — the "grandchildren" of the monarchs that flew south to Mexico the previous fall — usually arrive in central and eastern Ontario sometime in late May or early June. They'll be looking for milkweeds on which to lay their eggs, so try to assure you have a patch of a dozen or so of these plants somewhere in your yard.

Mid-June

- Bees gather by the thousands at the fragrant blossoms of black locust trees. Some even crawl about on the ground, too sated with nectar to be able to fly.

Late June

- Butterfly-watching is at its most productive in late June and early July, with the greatest number of different species in flight at this time. Tiger swallowtails, white admirals, and European skippers are especially noticeable. (See page 147)

- A meadow alive with fireflies is one of the most beautiful sights of late spring and early summer. Hovering like tiny helicopters, their flight projects an unhurried, calming quality. Actually a type of beetle, fireflies possess a special organ in the abdomen which produces light and serves to attract a mate. The light switches on when the insect flies upward and switches off when it descends. When a female of the same species sees the male's flash, she responds with her own luminous signal. Because different species of fireflies have different flash patterns, the female is attracted only by the proper sequence of flashes.

- Butterfly counts, which are similar to Christmas Bird Counts, take place across central and eastern Ontario in June and July. They provide a measure of how butterfly populations are faring.

Some Common Aquatic Invertebrates
(List organized by family)

Stonefly larva	Water scorpion
Mayfly larva	Giant water bug
Caddisfly larva (usually in case)	Water strider
Dobsonfly larva (hellgrammite)	Water boatman
Damselfly larva	Backswimmer
Dragonfly larva	Mosquito larva
Water penny larva	Black fly larva
Predaceous diving beetle	Leech
Whirlgig beetle	Crayfish

MOTH MONTH

Central and eastern Ontario is a veritable moth paradise. In fact, there are probably at least ten times as many moth species here as butterflies. Distinguishing between moths and butterflies is fairly straightforward. Butterflies usually perch with their wings held upward and have club-like knobs on the ends of the antennae. Moths, on the other hand, perch with their wings outspread and have antennae that closely resemble bird feathers. Unlike butterflies, most moths are nocturnal. Only a relative handful of moth species is active during the day. Among these, the blue-black, day-flying Virginia Ctenuchid moth is a common sight in fields.

June is the month of the spectacular giant silk moths. These impressive insects can measure up to 15 centimetres in width. The male silk moths have large, feather-like antennae which serve to locate a female. The female moths release airborne sex attractants called pheromones. These are emitted in infinitesimally small quantities.

However, the male's antennae are designed to maximize the surface area that can come into contact with the molecules. This explains why they are so much larger than the female's antennae. This amazing chemical communication system allows a male to find a female at distances of up to five kilometres!

Probably the best known of the silk moths is the Cecropia. After fertilization, the female Cecropia will lay 100 or more eggs, usually in a cherry, birch, or maple tree. The Cecropia caterpillar will reach nine centimetres in length by mid-August and then spin a spindle-shaped cocoon in which it will spend the winter. It will exit the cocoon as an adult in the spring. It is interesting to note that adult silk moths exist for the sole purpose of reproduction; in fact, they don't even eat. Some of the other silk moths that can be found in central and eastern Ontario include the Polyphemus, the Promethea, the Luna, and the small but spectacular Io moth.

Although silk moths may turn up just about anywhere, they are most common in rural areas. They are usually found from early to mid-June, although the Cecropia is sometimes seen in late May in years when warm weather arrives ahead of schedule. Silk moths are most active after 10:00 p.m. on warm, still nights. They are attracted to bright white lights. Look for them flying around the light, resting on the light pole, or sitting on the ground. Royal moths such as the imperial, and sphinx moths such as the big poplar can also be found at these same lights. Sometimes, a giant water bug will show up.

Unfortunately, our pervasive use of outdoor lighting may be harming Ontario's moth populations. Because of their nocturnal lifestyle, moths are very vulnerable to bright lights and especially those that are left on all night. When out on an egg-laying mission after dark, a female moth will often be pulled in by the allure of a light. Confused by all of the illumination, she eventually finds a nearby resting place where she will spend the night. Doom strikes at dawn when birds appear on the scene and recognize an easy breakfast. Whenever possible, we must assure that outdoor lights are not left on all night long.

The Cecropia, Canada's largest native moth, only lives for five to seven days.
Marvin Smith, Wikimedia

BUTTERFLY-WATCHING AT ITS BEST

One of the most pleasurable nature activities in summer is butterfly-watching. More species can be seen in late June and early July than at any other time of the year. Unlike birding, which often requires getting up at the crack of dawn, dealing with less than perfect weather, and straining your eyes and your neck muscles for a momentary flash of colour in the dense foliage, butterfly-watching is a much more civilized affair. Butterflies fly only during warm, sunny weather and are rarely on the wing before 8:00 a.m. As more and more people are discovering, butterflying can quickly become an obsession, especially if you also enjoy photographing them. Thanks to central and eastern Ontario's geographic location and mix of habitats typical of both more southern and northern regions of the province, it is possible to see over 100 butterfly species, with most occurring annually.

To find a given species of butterfly, it is necessary to know the time of year it flies and the kind of habitat it prefers. As a general rule, most species prefer sunny locations with lots of flowering plants. Joe-pye weed, milkweeds, and dogbanes are particularly attractive to butterflies. Rail-trails, roadsides, and the edges of wetlands and sedge meadows are often especially productive. Once you start paying attention, the abundance of some species, such as the European skipper and the northern crescent, is amazing. In addition to close-focusing binoculars and a guide book, you may also wish to take along a digital camera. By using the zoom feature, you can get close-up pictures showing the various field marks. In this way, you can wait until you get home to make a definitive identification.

Some Common Early-summer Butterflies

(List organized by family)

Canadian tiger swallowtail	Mourning cloak
Black swallowtail	American lady*
Cabbage white	Red admiral*
Mustard white	White admiral
Clouded sulphur	Viceroy
Acadian hairstreak	Northern pearly eye
Coral hairstreak	Eyed brown
Summer azure	Common wood-nymph
Great spangled fritillary	Monarch
Aphrodite fritillary	Columbine duskywing
Atlantis fritillary	Delaware skipper
Pearl crescent	European skipper
Northern crescent	Long dash
Eastern comma	Dun skipper

Numbers vary greatly from year to year

The Canadian tiger swallowtail is a common
butterfly species in the open woodlands.
Terry Carpenter

PLANTS AND FUNGI

JUNE HIGHLIGHTS

All Month Long

- More than 20 species of orchids bloom this month. Among them is the spectacular showy lady's slipper. (See page 150)

- With over 15 hours of sunlight daily, plant growth proceeds at amazing rates. White ash shoots may grow two centimetres a day.

- Fresh, immaculate June leaves cover the entire spectrum of green. The new bright lime growth on spruce trees is especially attractive.

- Early June through mid-July is the best time to check out the intriguing plants of bog habitats. Some of the most interesting species include bog laurel, bog rosemary, Labrador tea, leatherleaf, large cranberry, pitcher plant (carnivorous), round-leaved sundew (carnivorous), sheathed cottongrass (a sedge), and orchids such as grass pink and rose pogonia.

- A completely natural spring bloom of blue-green algae occurs in our lakes and is sometimes quite noticeable. The organisms feed on naturally occurring inorganic nutrients such as phosphorus and nitrogen brought into the lake by spring runoff. The species typically involved in spring is *Aphanizomenon flos-aquae*, a filamentous alga. Under some circumstances, it may be toxic.

- Many kinds of grasses bloom this month and fill the air with pollen. Among these are the forage grasses like orchard grass, Timothy, and meadow fescue.

- Prairie smoke blooms abundantly in alvar habitats such as the Carden Alvar at Kirkfield. Huge swaths of field are turned a sea of smoky pink. Indian paintbrush adds dabs of red, while balsam ragwort provides splashes of yellow.

- Once all of the leaves are out, we become aware of the many dead, leafless branches on a growing number of native trees. One of the biggest concerns right now is for the endangered butternut tree. It is being devastated by a fungal disease known as Butternut Canker.

Early June
- A large variety of wildflowers of coniferous and mixed woodlands is in flower. Most are pure white. (See page 151)

- The annual roadside flower parade begins with mustards and buttercups blooming first. They are followed later in the month by ox-eye daisies, dame's rocket, viper's bugloss, goat's beard, yellow hawkweed, and bladder campion. Almost all of the roadside denizens are exotic (non-native) species.

- The white flowers of hawthorns, dogwoods, and the first viburnums brighten fields and wetland edges.

Mid-June
- Balsam poplars, along with a variety of willows, release their airborne seeds. The seeds are carried great distances by long, silky, white hairs. Willow "fluff" collects on the surface of wetlands and is further dispersed by water.

- Serviceberries, also known as Juneberries, are the first shrubs to boast ripe fruit. The berries are a great favourite of birds such as American Robins.

- The male cones of pines and balsam fir release their pollen and powder both land and water with a yellow dust. (See page 152)

Late June
- Common St. John's wort is in bloom. Brought to North America from Europe, it got its name because it flowers in June and was traditionally harvested on St. John's Day, June 24. This, of course, coincides with the summer solstice. St. John's wort was thought to be imbued with the power of the sun.

- Common elderberries, along with narrow-leaved and broad-leaved cattails, bloom along wetland edges. On a cattail, the flower structure consists of a dense, dark brown, cylindrical spike on the end of a stem. The staminate (male), pollen-producing portion is located above the pistillate (female) portion.

- White water lily and yellow pond lily bloom. The fragrant flowers of the white water lily have a unique pollination method. On the first day the flower blooms, a fluid fills the centre of the flower and covers the female parts. Should an insect — usually a beetle — visit the flower, the petal design causes the insect to fall into the fluid. Any pollen on the insect's body dissolves in the fluid, causing the flower to be fertilized.

"FIRST BLOOM" CALENDAR FOR SELECTED TREES, SHRUBS, AND HERBACEOUS PLANTS

Early June: Bur oak, red oak, American beech, European buckthorn, starflower, wild sarsaparilla, goldthread, bunchberry, yellow clintonia, twinflower, Canada mayflower, Solomon's seal, prairie smoke.

Mid-June: White pine, black locust, black cherry, black walnut, alternate-leaved dogwood, highbush cranberry, nannyberry, bittersweet, poison ivy, red raspberry, blackberry, showy lady's slipper, Arethusa, grass pink, rose pogonia, wood lily, blue flag, blue-eyed grass, goat's beard, dame's rocket, bladder campion, forget-me-not, Canada anemone, yarrow, riverbank grape, yellow hawkweed, Philadelphia fleabane, ox-eye daisy, vetches, roses.

Late June: Catalpa, common elderberry, staghorn sumac, spotted coral root, common milkweed, viper's bugloss, white sweet clover, climbing bittersweet, rough cinquefoil, bird's foot trefoil, chicory, wood sorrel, sheep laurel, tall meadow rue, St. John's worts, cattails, yellow pond lily, white water lily, fireweed.

A TIME OF ORCHIDS

For many naturalists, June is synonymous with orchids. More than 20 species of these exceptionally beautiful wildflowers bloom this month in central and eastern Ontario. In addition to their legendary beauty, finding orchids is very satisfying because they usually require some special searching. Unfortunately, they have disappeared from many of their former locations, mostly because of habitat loss and degradation. The lady's slippers are probably the most renowned of our orchids, but they are by no means common. They have extremely complex flowers in which self-pollination is all but impossible. Bees enter the flower through the incurved split in the main pouch. They are forced to exit, however, by an opening at the top of the flower, and when they do so they inadvertently pick up pollen as they leave. If they visit another flower of the same species, they will follow the same path but unwittingly deposit pollen from the first flower, thereby assuring cross-pollination.

Central and eastern Ontario boasts five species of lady's slippers. Possibly the best known member of this genus is the pink. It is usually found in dry upland sites, almost always in association with pine. The largest of our native orchids is the showy lady's slipper. It measures up to 80 centimetres in height and occurs in open to semi-shaded wetland edges. Showy lady's slippers require ten years of growth from germination to the time they flower.

Dry to moist calcium-rich sites are the preferred habitat of the yellow lady's slipper. Some areas of central and eastern Ontario also have ram's-head lady's slipper, a species that has become quite rare provincially. This species prefers cold, undisturbed wetland edges, usually where white cedars are present.

Naturalists also seek out three other species of orchids this month because of the unique design of their flowers and the special habitats in which they grow. They are the dragon's mouth (Arethusa), grass pink (Calopogon), and rose pogonia. All three are usually found in wetlands such as fens and bogs, where they add beautiful splashes of pink to the landscape. These species are by no means easy to find, but the joy of discovery makes the search well worthwhile. The Bruce Peninsula is the premier orchid destination in Ontario.

The yellow lady's slipper has an inflated yellow pouch and yellow-green twisted petals.
Terry Carpenter

WILDFLOWERS OF THE CONIFEROUS FOREST

Just as the spring ephemerals of the hardwood forest are characteristic of May, the flowers of the coniferous and mixed forests typify the month of June. With the exception of stands of white cedar, habitats with a large proportion of coniferous trees are usually not as densely shaded as deciduous forests and the light conditions tend to be uniform over the entire year. Therefore, the plants associated with these habitats do not have the tight time constraints of the spring ephemerals to leaf out, blossom, and produce seed. They must, however, be tolerant of relatively low light conditions. Some species that are particularly fond of shade include fringed polygala, Canada mayflower, pink lady's slipper, wood sorrel, bunchberry, and wintergreen. In fact, the photosynthetic processes of these plants are most efficient at low light levels.

These plants share other characteristics as well. They are all perennials and most of them produce immaculate white flowers. In addition, species that thrive in poor light conditions often grow in colonies by means of a subterranean network of rootstalks. Bunchberry is actually an "underground shrub" in that the stems are all attached underground by a common rootstalk. This strategy allows the plant to dispense with an energy-demanding trunk and branch system. Rather than wasting the little sunlight energy it receives, bunchberry simply allows the ground to support the stems and leaves. Relatively few of the stems produce flowers, and the plant spreads mostly by underground rhizomes that eventually produce colonies of self-supporting clones. Canada mayflower, goldthread, wild sarsaparilla, and starflowers are just a few of the many plants that use this same strategy.

Unlike the spring ephemerals, the wildflowers of the coniferous forest have leaves that persist at least until the fall. Many, such as the wood sorrel and twinflower, have evergreen leaves that contain sugar compounds which act like antifreeze. They allow the leaves to survive the rigours of winter and to immediately begin photosynthesis in the spring.

A Dusting of Yellow

For a few brief days in late May or June, you may have noticed a mysterious dust that turns flat surfaces a lemon colour and makes the shores of lakes and streams appear as if a strange yellow alga has bloomed. This bizarre phenomenon is simply a manifestation of the sex lives of our coniferous trees as huge amounts of pollen are released to the wind every spring. When the weather is hot and dry, you can sometimes see what looks like a yellow shroud around certain conifers as the wind jostles the male cones on the boughs and spills copious amounts of pollen into the air.

All conifers produce cones. In fact, this is where the word *conifer* comes from. Cones are simply the reproductive parts of an ancient branch of plants known as gymnosperms. Gymnosperms are different from angiosperms such as maples or oaks in that they lack true flowers. The pollen grains land directly on the ovule, rather than on a flower structure like the stigma. As with many types of flowers, cones can be either male or female. The female cone consists of a central stalk surrounded by stiff, overlapping scales reminiscent of wooden shingles. If you open up a mature cone before it falls from the tree, you can often see the seeds inside. The male cones, also known as pollen cones, are a much smaller and far less conspicuous structure.

White pine is typical of many of our conifers with regard to pollination. The yellow-brown male cones are catkin-like in appearance and grow in clusters near the base of new shoots. They are usually located in the lower part of the crown of the tree. After the pollen is shed, these cones wither and fall away, often dropping from the trees like a veritable rain shower and covering the ground beneath the trees. The female cones become receptive to the wind-blown pollen at precisely the same time as the pollen grains are being shed. At this time, they are very soft and their scales are partially separated. A pollen grain is uniquely designed for wind pollination and actually contains two air bladders. They give it buoyancy and enable the pollen to take what amounts to a balloon ride. Following pollination, the scales on the female cones grow together. A pitch-like material then seals the outside. Over the next two years, the cone gradually grows to full size. In both red and white pine, the seeds are ripe by August or September of their second summer. The seeds are released to the wind when the cone scales open up. Pine are unique in taking two years from cone pollination to maturation and seed release. For spruce, fir, tamarack, cedar, and hemlock, all of these events happen over one year. The cones open and shed their seeds during the first fall or winter following pollination. It should be noted that the chemical composition of conifer pollen makes it less likely to produce allergic symptoms than other types of pollen.

WEATHER

JUNE HIGHLIGHTS

All Month Long

- With more than 15 ½ hours of sunlight over the entire month, June days convey a sense of unending time. Be sure to savour these long, sweet days!

- For all intents and purposes, June is considered to be a summer month. Weather is much more like summer than like spring. The daily mean temperature is only about three degrees cooler than July, our warmest month.

- The June air smells delicious. Keep your nose peeled for the scents of black locust blossoms, spirea, and the season's first freshly cut hay curing in the sun. It is claimed that from a downwind location you can smell a hayfield a mile away.

- Severe storms are commonplace in central and eastern Ontario, particularly in Simcoe County, where there is a relatively high prevalence of tornadoes. As a result of climate change, there will likely be a trend for more moisture in a warmer atmosphere. This is expected to cause an increase in extreme weather events such as windstorms. There are indications that this trend has already begun.[9] These storms do great damage to forests, as occurred in August 2006, when the Township of Galway, Cavendish, and Harvey in Peterborough County was hit by a severe windstorm and an F0 tornado.[10]

Late June

- Summer officially begins on or about June 21 with the summer solstice. The sun rises and sets farther north than on any other day of the year. It also stands highest in the sky at noon and casts the shortest shadows of the year. (See below)

- For the two middle weeks of the month, the sun rises at almost exactly the same time every day.

JUNE WEATHER AVERAGES, 1971–2000[11]

City or Town	Daily Max. (°C)	Daily Min. (°C)	Rainfall (mm)	Snowfall (cm)	Precipitation (mm)
Owen Sound	21.4	11.1	76.3	0	76.3
Huntsville	22.7	11.1	82.1	0	82.1
Barrie	23.4	12	86.6	0	86.6
Haliburton	22.1	9.8	86.8	0	86.8
Peterborough	23.5	11.7	78.9	0	78.9
Kingston	21.4	12.1	72.3	0	72.3
Ottawa	23.8	13	91.2	0	91.2

THE SUMMER SOLSTICE

A pivotal celestial event takes place this month. On or about June 21, we witness the summer solstice, the longest day of the year and the first official day of summer. At the solstice, Earth cruises past the point in its orbit that results in the greatest tilt of the Northern Hemisphere toward the sun. The sun rises and sets at its farthest point north and therefore traces its highest and longest arc through the sky. For several days before and after the summer solstice, the "sol" (Latin for sun) appears to "stice" (Latin for standing still) in the sky — that is, it rises in exactly the same spot on the northeast horizon and sets in precisely the same position on the northwest horizon. If you were to watch a time-lapse movie of a year's worth of sunrises, you would notice that the sun appears to "walk" back and forth across the western horizon. The winter solstice marks the southern limit of the sun's journey and the summer solstice is the northern boundary. At each end of the walk, the sun pauses for a few days before heading in the opposite direction. At the summer solstice, sunlight strikes our part of the globe more perpendicularly than at any other time of year and therefore heats the Earth much more efficiently. And all life responds! You have no doubt noticed the incredible amount of growth that occurs in your garden and on your shrubs and trees at this time of year.

Also known as Midsummer, Litha, or St. John's Day, the summer solstice used to be celebrated with astonishment, joy, and thankfulness by cultures all over the world. In 1996, the Canadian government declared June 21 as National Aboriginal Day in Canada because of the cultural significance of the summer solstice to aboriginal peoples. The solstice is best observed from a height of land that provides an unobstructed view of the northeast.

APPROXIMATE JUNE SUNRISE AND SUNSET TIMES (DST)[12]

(Note: Twilight starts about 30 minutes before sunrise and continues about 30 minutes after sunset.)

Location	Date	Sunrise (a.m.)	Sunset (p.m.)
Grey-Bruce	June 1	5:42	9:01
	June 15	5:38	9:10
	June 21 (solstice)	5:39	9:12
	June 30	5:42	9:13
Muskoka/Haliburton	June 1	5:38	8:55
	June 15	5:34	9:04
	June 21 (solstice)	5:29	9:06
	June 30	5:38	9:07
Kawartha Lakes	June 1	5:33	8:49
	June 15	5:29	8:59
	June (solstice)	5:29	9:01
	June 30	5:32	9:01
Kingston/Ottawa	June 1	5:26	8:42
	June 15	5:22	8:51
	June 21 (solstice)	5:22	8:53
	June 30	5:25	8:54

NIGHT SKY

JUNE HIGHLIGHTS

All Month Long
- Major constellations and stars visible (June15 — 10:00 p.m. DST)
  Northwest: Ursa Major high in sky; Ursa Minor (with *Polaris*) to its right; Leo (with *Regulus*) to left in mid-sky.
  Northeast: "Summer Triangle" made up of brightest stars *Vega* (in Lyra), *Deneb* (in Cygnus), and *Altair* (in Aquila); Cassiopeia low in the north.

Southeast: Bootes (with *Arcturus*) high in sky; Sagittarius at eastern horizon and, above it, the brightest part of the Milky Way.
Southwest: Virgo (with *Spica*).

- Starting in June, the Milky Way is at its most spectacular. This is because we are now facing our galaxy's densely star-studded centre. (See below)

- The Algonquian name for the full moon of June is the Strawberry Moon.

- The summer stars have arrived. The three stars of the Summer Triangle — Vega, Deneb, and Altair — can be seen low in the eastern sky soon after dark.

- Look high overhead for Arcturus, the star that heralded the arrival of spring and now the brightest star in the sky.

- The June moon rises about 30 degrees south of due east and sets 30 degrees south of due west.

THE MILKY WAY

Anyone with an interest in the night sky looks forward to the period from late spring to early fall for a veritable rite of summer stargazing — observing the Milky Way. It is the most beautiful feature of the night sky during these warm nights and should not be missed. Shaped like a spiral wheel, the Milky Way is our home galaxy. Our solar system is located on one of the spiral arms about two-thirds or 25,000 light years out from the centre. The Milky Way is brightest in the area of the constellation Sagittarius, which lies in the southeast. This is because the centre of the galaxy is actually located behind the stars of Sagittarius. There are also dazzling parts of the Milky Way in Cygnus and in Perseus. For optimal viewing, try to choose a clear, moonless night when the temperature gets down to at least 15°C. Warm, humid nights tend to produce a lot of haze. There is no need to have a telescope. A pair of binoculars will transform the gauzy "river of milk" seen by the naked eye into thousands and thousands of individual stars. Viewing is at its best from 11:00 p.m. to 1:00 a.m.

JULY

Summer at Its Height

QUEEN ANNE'S LACE.

BLACK-EYED SUSAN

WHITE SWEET CLOVER

Some common roadside flowers of July.
Kim Caldwell

Live in each season as it passes:
breathe the air, drink the drink, taste the fruit.

— Henry David Thoreau

THE PREPARATIONS AND TOIL OF SPRING BEGIN TO BEAR FRUIT IN JULY. EGGS ARE now fledged birds; the early flowers have become ripe berries; tadpoles have grown into small frogs; and the once-green roadsides are now a riot of colour and floral diversity. Where several weeks ago there were mostly whites and yellows, the intense colours of high summer have added blues, oranges, pinks, and magentas to the roadside palette.

July is a treat for all of our senses. The warm, humid air is often replete with the sweet smell of milkweed flowers, while the fragrance of flowering basswood trees draws bees and other insects by the thousands. Our palates, too, are well-served in July as strawberries, raspberries, tomatoes, and the first sweet corn ripen. Although the annual cycle of birdsong is winding down now, the serene, haunting song of the thrush gives beauty to early summer evenings. In some areas, the voice of the Whip-poor-will still rings out as darkness falls.

This, the warmest month of the year, brings hot, humid, and thundery weather. Afternoons shake and tremble with intense thunderstorms whose gift of rain is often too short and violent to be of much benefit to the thirsty soil. The fact that the sun is rising and setting a little farther to the south each day largely goes unnoticed as we go about enjoying the summer weather. But like a cruel joke, the first thing that happens once true summer has finally gotten here is fall migration. The vanguard of southward-bound shorebirds is already starting to arrive in central and eastern Ontario from the Arctic and by month's end the first warblers will be departing. And so the wheel of the year continues to turn, allowing us very little time to simply sit back and admire the fresh, new world around us — the intimations of autumn are already beginning to make themselves known.

BIRDS

JULY HIGHLIGHTS

All Month Long
- Flocks of post-breeding European Starlings, Red-winged Blackbirds, Common Grackles, and American Crows are noisy and conspicuous. (See page 160)

- Ruby-throated Hummingbirds are a constant source of wonder and delight as they visit our gardens and feeders all summer long. (See page 161)

- All summer, the familiar birds on cottage country lakes can include Common Loons, Herring Gulls, Caspian Terns, and, increasingly, the Double-crested Cormorant.

- Family groups of Common Mergansers are sometimes seen feeding and travelling along shorelines on lakes in Shield country. Because broods of mergansers sometimes combine, it is not uncommon to see a female with a parade of over 20 young in tow.

The mother protects the chicks, but she does not feed them. They dive to catch all of their own food.

- American Goldfinches and Cedar Waxwings breed later than most other birds. Goldfinches usually wait until sometime in July when thistles have produced their downy seeds. Not only is thistledown used to line the nest but regurgitated thistle seeds are fed to the young. Waxwings nest any time between early June and early August as berry crops, their main source of food, begin to ripen.

- Cottage roads can be good for birding during the early summer. Watch especially for trees and shrubs with ripe fruit where birds may be feeding and for brushy areas which afford lots of cover. If you hear contact calls, stop and pish. It is often possible to attract large numbers of songbirds — including many warblers — in this manner. Some may have recently fledged young in tow.

Early July

- As the nesting seasons wraps up, there is a marked decrease in birdsong. Singing requires a huge amount of energy, so if there is no reproductive imperative to do so, it is to a bird's advantage to remain quiet. In many ways, this is the turning point of the avian year, since "fall" migration for some species has already begun.

- With thoughts of warm Latin American beaches already on their minds, the first southward-bound shorebirds start appearing at sewage lagoons and at beaches on the Great Lakes. These early migrants are individuals who have failed to nest successfully on their Arctic breeding grounds. There are also females of many species (e.g., Lesser Yellowlegs) that simply leave the male with the young and head south!

- By now, young Bald Eagles are usually able to make their first solo flights from the nest. They will, however, remain in the vicinity of the nest for most of the summer.

- Veeries, Wood Thrushes, and Hermit Thrushes still sing their clear, serene songs on summer evenings. Hermit Thrushes often sing during the day, too.

- The high, paired notes of the Indigo Bunting ring out from telephone wires and tree-tops on the margins of shrubby fields. The bright blue male is an inveterate songster, singing even during the hottest part of the day.

Mid-July

- Caspian Terns, the world's largest tern, start showing up on many cottage country lakes as they disperse from breeding colonies on Lake Huron and Lake Ontario.

- Swallows start congregating on roadside wires, especially in the vicinity of farms. Tree and Barn Swallows are the dominant species.

- Bobolinks flock up and leave their grassland and hayfield breeding habitat. They often move into marshes to moult before migration.

Late July

- Northern Waterthrushes and Yellow Warblers that did not successfully raise young begin to migrate south.

- By the end of the month, shorebird-watching is usually quite good at local sewage lagoons and at beaches on the Great Lakes such as Presqu'ile Provincial Park.

- Young Ospreys leave the nest in late July or early August but will still return to roost at night for about a month. During the day, they usually perch near the nest, and make near-constant food calls. The male still provides most of their food. The young won't catch fish on their own until one or two months after starting to fly.

Despite its unassuming appearance, the Hermit Thrush's flute-like, haunting song is one of the most beautiful of any Canadian bird.

Karl Egressy

POST-BREEDING FLOCKS

Already, a foretaste of autumn bird behaviour can be seen in some bird communities. Several species are already assembling in flocks as males begin to lose their intolerance of one another. Starting in July, blackbird flocks become a common sight in both urban and rural areas. In wetland habitats, small flocks of Red-winged Blackbirds and

Common Grackles can be seen. Right now, they are fairly small and only include the adults and young of local breeding populations. Swallows, too, begin to gather in post-breeding flocks this month and often roost at night in marshes.

For city dwellers, European Starlings and American Crows are especially noticeable because of their roosting (gathering together to spend the night) behaviours. Deciduous shade trees, which are prevalent in urban and suburban areas, are the preferred roosting sites. Hundreds of crows or starlings may occupy a given stand of trees and, unfortunately, sometimes continue to return each night until the leaves drop. As sunset approaches, the birds start arriving in the vicinity of the roost and perch in nearby trees, often making frequent, clamorous flights from one tree to another. This activity, known as staging, goes on for about half an hour before they actually settle into the roost trees.

The extent to which crows are both roosting and nesting in towns and cities is a relatively new phenomenon. The birds seem to have figured out that it provides more advantages than disadvantages. Although crows can be legally hunted, it is illegal to discharge firearms in urban areas. Cities are also appreciably warmer than rural areas because of the so-called "heat bubble" that surrounds them; artificial lighting such as street lights helps crows watch for predators such as the Great Horned Owl, which they are hard-wired to fear; and, in some areas, cities and towns provide the largest and best roosting and nesting trees.

Forming flocks, too, provides advantages to birds. First of all, there is safety in numbers. The chance of any one individual being killed by an enemy is lower in a flock than if that bird was by itself. When predators attack a flock, they try to single out a bird on the edge of the group and pursue that one individual. However, most flocks change shape constantly, expand and contract in size, and generally make it very difficult for the predator to remain focused on one bird. Birds also appear to gain information about good feeding sources by following other birds in the morning when they leave the roost. While the birds are eating, it takes only a few individuals to watch for danger. That allows the vast majority of the birds in the flock to simply focus on feeding and preening.

AMAZING HUMMINGBIRDS

When it comes to summer guests in cottage country, very few are more engaging than the Ruby-throated Hummingbird. Its tiny size, agility of flight, pugnacious behaviour, and iridescent colours are only a few of the characteristics that make this tropical migrant so special. With an average of 40 to 80 wingbeats per second, some incredible aerial manoeuvres are possible. In addition to being able to hover, hummingbirds can fly sideways, up, down, forward, and even backward. It is also hard not to be impressed that they make a beeline from their wintering grounds in Central America to the same summer territory and feeder year after year.

Hummingbirds have enormous appetites and need to refill their stomachs about every seven to 12 minutes. This means consuming food equal to about half their body weight each day. As Doug Bennet and Tim Tiner explain in their excellent book *Wild City*, in order to sustain the metabolic rate of a hummingbird, "a 176-pound man would have to pack away 100 pounds of Smarties a day and drink five cases of beer to keep his skin from catching fire."[1] With all this frenzied feeding activity, you may wonder then why the birds spend so much time sitting quietly on a perch, seemingly doing nothing. They are in fact going through the process of emptying their crops. Before being able to feed again, the hummingbird has to wait for its crop to become about half empty, as sugar water or nectar is passed into the rest of the digestive system. Hummingbirds aggressively defend their food supply against intruders, including other hummingbirds, and often employ special flight patterns in displays of aggression. In one such display, the bird describes a horizontal *U*, going back and forth and passing close to the intruder's ears. Both males and females engage in this behaviour.

The number of hummingbirds coming to feeders usually increases in July, when fledged young begin to accompany the female on feeding excursions. They are almost identical to their mother in appearance. Finally, the hard-working female has some time to catch her breath, having built the nest, incubated the eggs, and taken full responsibility for feeding the nestlings — all without any help from the male. To feed hummingbirds, use a solution of one part sugar to four parts water. The mixture should be changed at least once a week.

MAMMALS

JULY HIGHLIGHTS

All Month Long

- Bears sometimes come into rural neighbourhoods and cottage communities in search of food. (See page 163)

- Roadkill on our highways is very noticeable. The carnage is partly due to the large number of young mammals that must range widely in search of food and/or a new territory. Simply slowing down is often enough to avoid many collisions with wildlife.

- Bats are seen more often since the young are now starting to fly. In fact, a mother and her youngsters are often seen flying together. Until they can catch their own food, the youngsters are entirely dependent on their mother's milk.

- Throughout the summer, elk can sometimes be seen grazing in cattle pastures and hayfields. However, by doing so, they are consuming food intended for cattle, which leads to conflict with farmers. Between 2000 and 2001, 120 elk were released in the

Bancroft area. The animals originated from Elk Island National Park in Alberta. In 2010, the population was estimated to be between 330 and 766 animals.[2] They are sometimes seen in the Hartsmere Road area at the north end of Weslemkoon Lake, east of Bancroft.

Mid-July
- Mammals such as bears, raccoons, and foxes stuff themselves as blueberries, raspberries, and cherries ripen by the middle of the month. The large berry component of mammalian diet makes the scat (droppings) almost impossible to identify at this time of the year. It appears in all sorts of shapes and configurations.

BEARS AND PEOPLE[3]

There is a general perception among many people that black bear numbers have increased significantly in recent years. But according to Bear Wise, a MNR website devoted to information on black bears, it is the increase in the human population, particularly in cottage country, that is a more credible explanation for more contact between people and bears. This does not necessarily suggest more bears. There is, however, good evidence that the bear population has been slowly growing over the past 20 to 30 years along the southern edge of the Canadian Shield. As abandoned farmland returns to old field habitat, it provides much better habitat for bears. Raspberries, blueberries, aspen, and hawthorn all grow in this habitat and provide food and cover.

Encounters with bears are most common in years when natural food is in low supply. Many preferred foods of black bears are only available for short periods or vary greatly in availability from year to year. Late spring frosts, cool and wet springs or summers, drought and fire can all affect the supply of natural food for black bears. Even during an average year, most shrub and tree-borne fruit and nuts won't become available until mid-July. Since bears need to double their weight before going into winter hibernation, it is not surprising that just about everything these animals do is related to getting food.

Bear conflicts with humans generally arise as a result of improperly stored food and garbage or from feeding bears on purpose. Bears quickly learn to associate human residences and campsites with readily available nourishment, especially if garbage or other forms of food are left outside. You can greatly reduce the chances of a visit from a bear by doing things such as waiting until the morning of garbage day to put the garbage out, keeping your barbecue grill clean, and putting wild bird feed out only in the winter. Once bears learn to access human food, management options — besides destruction of the bear — are limited. If bears are successful at getting food, they will return again and again. If we want to keep bears wild, we need to ensure they do not lose their natural fear of people. The most typical encounter is the one you didn't know happened. The bear heard or smelled you, and left. However, should you ever encounter a bear, slowly back away while watching the animal and waiting for it to leave. To report bear problems, call the Bear Reporting Line at: 1-866-514-2327. In a life-threatening emergency, call your local police or 911. For more information, go to the Bear Wise website at *bears.mnr.gov.on.ca*.

Black bears lose weight through the spring and early summer until large quantities of berries become available.
Terry Carpenter

Amphibians and Reptiles

July Highlights

All Month Long

- Unlike most snakes, the northern water snake is curious and sometimes unwary of humans. Although it is harmless and will only bite in self-defence, water snakes have been known to approach swimmers. They may be doing this to see if a prey item such as a frog is actually the source of the ripples. Sadly, some people feel threatened by any snake they see and are quick to kill the animal. This is one of many reasons why, like turtles, Ontario's snake populations are in a serious decline. (See page 165)

- Clamorous frog song is but a memory of spring. However, green frogs, bullfrogs, and some gray tree frogs will continue to call until late July or early August, marking the end of the amphibian chorus.

- Young frogs transform into adults and leave their natal ponds. Tiny wood frogs, spring peepers, and American toads can usually be found in moist areas of the forest floor from July through September. However, the tadpoles of green frogs and bullfrogs must over-winter at least one year and often two years before they reach the adult stage.

- Salamander larvae also mature into the adult phase in July or August. Mole salamanders such as the spotted and blue-spotted take up residence in subterranean retreats such as beneath large rotting logs.

- Snapping turtles continue to lay eggs into early July, particularly during cool summers.

- Watch for the uncommon map turtle basking on rocks in large bodies of water such as Stony Lake, Georgian Bay, and the Ottawa River. They often do their sunbathing piled one on top of the other. Map turtles frequent deep water and are very hard to approach.

- Eastern garter snakes and northern water snakes give birth in July and August. The young of both these species are born live, since the eggs are retained within the mother's body until they are ready to hatch. There may be as many as 50 in a brood.

AN UNCERTAIN FUTURE FOR SNAKES

As with turtles, central and eastern Ontario's snakes and one species of lizard are facing an uncertain future. The region is home to 14 snake species, almost all of which have severely declined over the last century. Eight snake species and our one lizard species have been classified by the Ontario Ministry of Natural Resources (OMNR) as Species at Risk.[4] Of these, six are designated as Threatened — eastern hog-nosed snake, eastern fox snake (Georgian Bay population), black rat snake (Frontenac Arch population), queensnake, massasauga rattlesnake, and Butler's garter snake — and two are of Special Concern — eastern ribbon snake and milk snake. The southern Shield population of the five-lined skink is also of Special Concern.

Ontario's snakes face a plethora of challenges. However, habitat loss is the greatest single concern. The southern edge of the region, where the greatest number of species lives, has been dramatically altered by human activities such as agriculture and urbanization. Many species are now restricted to islands of habitat amid an ocean of human development. Snakes are also very vulnerable to being run over on roads, because snakes don't simply cross the roads but use them as a place to soak up sun and the heat from the pavement.

Sadly, some people have a deep-rooted, psychological aversion to snakes and probably can do little to change the way they feel. That is fine. One doesn't have to like snakes in order to respect them. What is most unfortunate, however, is that some individuals will intentionally harass or even kill these animals because of the misinformed belief that they are dangerous, evil, or simply useless. This is not only indefensible but also ecologically unsound, given the important role snakes play in the food chain. The milk snake often pays a particularly heavy price because of its habit of vibrating its tail. Because the vibrating tail can make a buzzing sound in dry leaves, this species is sometimes mistaken for a rattlesnake. It is also illegal to kill, harm, harass, capture, or take a living member of a species that is listed on the Species at Risk in Ontario list as an extirpated, endangered, or threatened species.[5] It's important to speak up when you see or hear of people persecuting snakes. Putting an end to such regrettable behaviour would be a major step forward in snake conservation. Many mammals, such as raccoons, skunks, and cats, also prey upon snakes. This is another good reason to keep your cat indoors and to never feed raccoons.

In the spring of 2009, the Ontario Reptile and Amphibian Atlas was launched in order to collect urgently needed information to monitor trends in species' distributions and abundance, assess population status, and identify and manage important habitat for rare species.[6] Another of the main objectives of the new atlas project is to collect observation submissions from the general public. The project wants to receive observations — even from past years — of all reptile and amphibian species, not just the rare ones. Observations can be submitted via an online form at *www.ontarionature.org*.

To defend itself against predators, the beautifully marked eastern milk snake sometimes vibrates its tail. However, it is harmless.

Joe Crowley

JULY | *Summer at Its Height*

FISH

JULY HIGHLIGHTS

All Month Long

- Largemouth Bass feed voraciously in shallow, weedy bays. (See below)

- Rock Bass, Pumpkinseeds, Bluegills, Yellow Perch, and young Smallmouth and Largemouth Bass provide fish-watching opportunities throughout the summer. Look for them in the vicinity of docks, vegetation, and other shoreline structures.

- Many species such as Brook Trout and Lake Trout are moving toward deeper waters as the lake temperatures rise as a result of warm summer weather.

- Large schools of Spottail Shiners can be seen over sandy lake bottoms. They are a favourite prey of many game fish and a commonly used bait fish. Spottails can be easily identified by the large black spot at the base of the tail.

- Fish die-offs occur with some regularity in summer and often involve carp. Die-offs can be caused by pathogens such as bacteria and viruses. When the fish are already stressed by factors such as abnormal water temperatures (when water becomes too warm, the oxygen content drops), storm events, spawning stress, and high population levels, even naturally occurring bacteria such as *Flavobacterium columnare* can spread quickly through fish populations and cause many deaths. A recent die-off in the summer of 2007 in the Kawartha Lakes killed several tens of thousands of Common Carp.[7]

SUMMER BASS

Largemouth Bass frequent shallow, weedy waters on July mornings and evenings, where they feed voraciously on prey such as frogs. This behaviour corresponds with the peak frog numbers, since young leopard frogs are just transforming into adults. Bullfrogs are also vulnerable in early summer, since their mating behaviour can attract the attention of large bass and Muskies. For many people, their best fishing memories are of warm, calm summer evenings casting for bass with a surface lure that mimics a frog. Smallmouth Bass, too, are popular with anglers. It is famous for its superb fighting abilities and its readiness to strike at bait throughout the summer months. Smallmouths prefer deep, cold water in July but will move up into shallower water after dark, where they can feed on nocturnal prey such as crayfish.

INSECTS AND OTHER INVERTEBRATES

JULY HIGHLIGHTS

All Month Long

- With so many different dragonflies and damselflies on the wing this month, identifying and photographing these beautiful creatures is a wonderful way to spend a summer afternoon. (See page 169)

- Deer flies, horse flies, and stable flies are a bothersome presence throughout July. (See page 171)

- Although warm water makes for excellent swimming conditions, you may need to be careful not to cut your feet on zebra mussel shells. This invasive species is now abundant in many lakes of central and eastern Ontario. (See page 171)

- Tiny, moth-like butterflies called skippers are by far the most abundant butterflies in many areas this month. They get their name from their fast, erratic style of flight. Watch especially for the European skipper, which is easily identified by its brassy, burnt-orange wings.

- If you're out for an early morning walk alongside a shrubby field or wetland, watch for large, circular spider webs called orbs. Constructed vertically to the ground, a big orb web can be half a metre in diameter and have 15 or more "spokes" reaching to the hub where the owner usually sits.

- Most of the tiny, delicate flies attracted to cottage and campground lights at night are midges. They resemble mosquitoes but lack the biting mouth parts.

- The tent caterpillars that were so prevalent on the cherry trees in May have now pupated and become small brown moths.

- Water striders and whirligig beetles are a common sight on the surface of rivers, ponds, and lakes all summer long. Being true bugs, striders impale their prey with their beak and then suck out the contents. They are able to actually walk on the water, thanks to the water-repellent hairs on their feet. Whirligig beetles, on the other hand, whirl around on the surface unmolested by fish because of their offensive smell. Abdominal glands produce an array of defensive chemicals, some of which smell like rotting diapers!

- Cottagers often see dock spiders (*Dolomedes scriptus*) in summer. Some individuals can be as wide as a CD, making them Ontario's largest spider species. They do not build a web but actively hunt at night for insects. However, they will even go below the surface on occasion in the pursuit of small fish. Air trapped in body hair allows them to breathe underwater for up to 45 minutes. Dock spiders can deliver a painful bite if handled.

- On summer days, hummingbird moths (family Sphingidae) such as the hummingbird clearwing put on a show to rival that of their avian namesake. Like hummingbirds, these fast-flying, clear-winged moths hover over flowers while feeding on the nectar with their long proboscis. At first glance, it's hard to believe you're actually looking at an insect and not a bird.

Mid-July

- The buzzy, electric song of the cicada is beginning now to fill the void created by the decrease in birdsong.

- Grasshoppers and crickets are suddenly very noticeable.

GETTING TO KNOW OUR DRAGONFLIES AND DAMSELFLIES

It is hard to go anywhere near water in July and not notice dragonflies and damselflies. Some even turn up in suburban gardens. To tell them apart, remember that dragonflies have thick bodies, are strong fliers, and their wings are open at rest. Damselflies are usually much smaller, have thin bodies, are weak fliers, and their wings are closed or only partially spread at rest. Some of the most abundant damselflies are powder blue in colour, hence the common name of "bluets." Belonging to an ancient order of insects called Odonata, we are seeing life as it existed millions of years ago. Because of their relative abundance and the fact that most can be readily identified in the field, they are an attractive insect group to study. As with butterflies, close-focusing binoculars and even a digital camera are useful for identification purposes. A butterfly net, too, can come in handy since some species must be captured for accurate identification. 172 species of dragonflies and damselflies have been recorded in Ontario.[8]

One of the most interesting and noticeable Odonata behaviours is mating. The extraordinary acrobatics that dragonflies and damselflies go through are definitely worth watching. There are three stages of Odonate sex. First of all, the male bends his abdomen beneath him in order to transfer sperm from his genitalia at the tip of the abdomen to the secondary sex organs located near where the abdomen and thorax meet. Secondly, the male forms a "tandem" with the female by grabbing her behind the head with claspers located at the tip of the abdomen. Finally, the pair alights and forms a "wheel" position. The female bends the tip of her abdomen around until her genitalia (located at the tip) are brought into contact with the male's secondary sex organs. He then may use special "scoopers" to clear out any sperm that another male may have deposited in the female. This helps to assure that only his genes will be transferred to future generations. Having cleaned house, the male injects his sperm into the female, and the wheel position is broken. Because the male wants to make sure that a rival suitor does not impregnate the female, males of most species will stay around and actively guard her until she has finished laying her eggs. In some species, including most of the damselflies, the male actually retains the female in his hold until egg-laying is complete. The *Field Guide to the Dragonflies and Damselflies of Algonquin Provincial Park and the Surrounding Area* is an excellent resource for anyone interested in central and eastern Ontario's Odonata.

Some Common Summer Dragonflies and Damselflies

(List organized by family)

Dragonflies	**Damselflies**
Common green darner	Ebony jewelwing
Canada darner	Common spreadwing
Lancet clubtail	Spotted spreadwing
Racket-tailed emerald	Marsh bluet
Dot-tailed whiteface	Hagen's bluet
Common whitetail	Familiar bluet
Chalk-fronted skimmer	Eastern forktail
Four-spotted skimmer	Sedge sprite
Twelve-spotted skimmer	
Widow skimmer	
Calico pennant	

The widow skimmer is often seen far from water, in fields, meadows, and gardens.
Terry Carpenter

DEER, HORSE, AND STABLE FLIES

Just as May is infamous for its black flies and June for its mosquitoes, July's most noticeable biting insects are the notorious deer and horse flies, both members of the Tabanidae family. Unfortunately, repellents that work well against mosquitoes and black flies are not as effective against deer and horse flies, since these big-eyed, day-time biters rely mostly on visual clues to find hosts and home in on tender flesh. Deer flies (genus *Chrysops*) are actually quite beautiful and can be quickly identified by their black-patterned wings, colourful bodies and iridescent eyes. They tend to buzz most persistently around the victim's head and shoulders. It is believed that deer are the ancestral host of deer flies. As with humans, the flies attack deer high on the body. We can at least be grateful that only the females bite. Nectar is the food of choice of the seldom-seen male. The name "horse fly" is generally used to designate Tabanidae flies other than deer flies. Most are in the genera *Hybomitra* and *Tabanus*. They are larger, grey or black in colour, and have huge eyes. Horse flies tend to bite lower on the body, preferring the legs. They are less common and generally not a serious pest to humans.

Stable flies (*Stomoxys calcitrans*) are members of the Muscidae family. They are about the size of a house fly — another name is "biting house fly" — but have four dark stripes on their thorax. They often attack the ankles and can inflict a rather painful bite with the long, pointed proboscis with which they suck blood. Stable flies frequently lay eggs in piles of rotting vegetation along shorelines and are regularly seen at beaches and cottage docks. As their name implies, they are also very common around barns and often bite livestock. Stable flies are fast-flying and much more difficult to swat than deer or horse flies.

ZEBRA MUSSELS

Zebra mussels, small clam-like shellfish of Eurasian origin, have been steadily increasing in central and eastern Ontario since the mid-1990s. Like dozens of other non-native species, they were first introduced to the Great Lakes in the ballast water of ocean-going ships. Their yellowish-brown, D-shaped shells can grow up to five centimetres in length and are sharp enough to tear skin. Zebra mussels attach themselves to solid objects such as submerged rocks and dock pilings in shallow, algae-rich water. Being filter-feeders, they are able to make a remarkable difference in water clarity. This allows for greater light penetration, which can in turn impact the entire ecological balance of the lake. Mature female mussels can produce in excess of 40,000 eggs per year. The eggs hatch into free-swimming, microscopic larvae known as veligers. This larval stage accounts for the rapid spread of the mussels. They are able to survive in any residual water source and are often inadvertently introduced into another lake. Veligers drift with the currents for three or four weeks as they seek out a hard surface on which to attach themselves. They then transform into the typical clam-shaped mussel.

Diving ducks such as Greater and Lesser Scaup eat large quantities of zebra mussels, but do not significantly reduce their population levels. Because the mussels concentrate contaminants in their tissues, the contaminants get passed on to the ducks in higher levels than when the birds eat other foods. This may have a negative impact on the ducks' survival and/or reproductive success. In some areas, populations of native clams have significantly decreased or completely disappeared as a result of zebra mussel infestations. The mussels will grow on top of the clams, preventing them from opening and closing properly. This also makes breathing and feeding difficult.[9]

PLANTS AND FUNGI

JULY HIGHLIGHTS

All Month Long

- Roadside flowers are at their most colourful and diverse. Ox-eye daisies usually dominate early in the month. (See page 174)

- A plethora of exotic invasive plants become very noticeable once early summer arrives. An important first step toward protecting our native botanical heritage is to familiarize ourselves with at least the most threatening species and to remove them where possible. (See page 175)

- Fireweed blossoms turn fields, forest edges, and burned-over areas a riot of pink. The flowers first start to open at the bottom of the stem.

- In our woodlands, particularly where conifers such as hemlock are present, a few wildflowers are still blooming. Some of the species to watch for include wood sorrel, twinflower, pipsissewa, wintergreen, and shinleaf.

- A profusion of ferns adorn fields, wetlands, and forests this month. By this time, they usually show all of their key characteristics and can therefore be identified more easily.

- In cottage country, watch for bolete mushrooms. Although they are the same size and shape as other mushrooms, they have no gills under the cap, but rather an underside surface full of tiny pores. Summer — not fall, as some people think — produces the greatest variety of mushrooms. Rainfall, along with several days of high humidity, is necessary to initiate fruiting.

Early July

- Prairie species such as butterfly milkweed and wild bergamot are flowering. Tallgrass prairie once covered nearly 1,000 square kilometres of central and southern Ontario but, due to agriculture and urbanization, less than 3 percent remains. (See page 176)

- Common milkweed is in bloom and its sweet scent fills the early summer air. The smell serves to attract insects so that their feet will inadvertently pick up the flowers' sticky pollinia — small packets containing pollen — and transfer them to another plant. If the insect is not strong enough, however, it can actually become stuck to the flower and die.

- Along cottage roads, watch for thickets of purple-flowering raspberry. A small shrub, it has maple-like leaves and rose-purple flowers about three to five centimetres across. The raspberry-like fruits have a nice flavour but tend to be seedy.

- Several species of orchids are still blooming, including rose pogonia, a wetland plant.

- By July, many of the flowering shrubs of May and early June are yielding ripe fruit. These include the crimson berries of the red elderberry. On most plants, hungry birds will have devoured them all by mid-month.

Mid-July

- Wetlands deliver a full spectrum of colour as a variety of shrubs and herbaceous plants are now in bloom. (See page 177)

- American basswood trees flower. The fragrant blossoms are rich with nectar and therefore attract hordes of bees and other insects. Together, they produce an extraordinary humming sound as the tree literally vibrates with life.

- Wild raspberries, blackberries, and blueberries are usually ripe and ready to be enjoyed.

Late July

- The branches of cherries, honeysuckles, and dogwoods bow over with ripe berries. Many bird and mammal species gorge themselves accordingly.

- Ghostly white Indian pipe blooms in the heavy shade of hardwood forests. It is one of the few woodland plants still in flower. It is leafless and contains no chlorophyll. Indian pipe receives its nourishment from fungi in the soil, which in turn draw their food from the roots of trees. Another parasitic, summer-blooming plant that shares similar habitat is the spotted coral root, a member of the orchid family.

"First Bloom" Calendar for Selected Trees, Shrubs, and Herbaceous Plants

Early July: Swamp milkweed, Joe-pye weed, harebell, black-eyed Susan, Queen Anne's lace, early goldenrod, wild bergamot, butterfly milkweed, hairy beardtongue, orange hawkweed

Mid-July: Evening primrose, bouncing bet, spotted jewelweed, spotted coral root, pearly everlasting

Late July: Showy tick-trefoil, boneset, cardinal flower, Kalm's lobelia, Indian pipe, rattlesnake plantains, grass-leaved goldenrod

The sensitive fern grows in wet woods. It is, as the name implies, very sensitive to frost, which can quickly kill the fronds.

Drew Monkman

The Roadside Extravaganza

July roadsides are bordered with numerous brightly coloured flowers. The majority of these species are non-native perennials. Unlike woodland species, their bloom period is quite long, usually averaging around 45 days. Some plants, like yarrow and butter-and-eggs, can actually produce flowers for more than 100 days — from June through to October.

Ox-eye daisies dominate early in the month, accompanied by an assortment of other species, including purple vetch, orange hawkweed, viper's bugloss, bladder campion, spreading dogbane, common milkweed, and Philadelphia fleabane. Toward the middle of the month, white sweet clover is usually the predominant plant. Also usually present in large numbers are Queen Anne's lace, wild parsnip, black-eyed Susan, purple-flowering raspberry, yarrow, and various cinquefoils and St. John's worts. By the end of July, Queen Anne's lace will have taken over as the predominant roadside flower. Some new plants have become much more conspicuous. These include bouncing bet, smooth hawk's beard, musk mallow, chicory, evening primrose, common mullein, fireweed, various thistles, knapweeds, and the first goldenrod. Pearly everlasting is also flowering in more northern regions in late July. In alvar and remnant tallgrass prairie habitats, wild bergamot is in bloom.

INVASIVE PLANTS

Not all non-native plants are created equal. Some, like ox-eye daisy, are generally harmless and coexist peacefully with the native vegetation. Many others, however, are what we call invasive species. They have come armed with aggressive reproductive qualities that enable them to displace our native plant communities. These so-called invasive exotics spread and survive so effectively that they choke out native plants. This, in turn, degrades habitats and can greatly reduce biodiversity. Invasive exotic plants tend to mature quickly, produce enormous amounts of seed every year, and establish themselves easily on disturbed sites. They can also be very difficult to remove or control. If at all possible, they should be cut back before seeds are produced. Here are four of the most threatening.

- Also known as swallow-wort, dog-strangling vine is a perennial with small purplish flowers. A member of the milkweed family, the plant produces narrow seedpods which taper to a slender tip. The pods release small brownish seeds with long, silky white parachutes. Dog-strangling vine grows in nearly all types of open and semi-open habitats, including the understory of woodlots. It sprawls and twines over fences and other vegetation and often forms impenetrable masses, hence its "strangling" moniker.[10]

- Ontario woodlands and fencerows are also under siege by a rather gentle looking flower known as garlic mustard. Named for the garlic fragrance produced when the leaves are crushed, it grows in both dense shade and sunny sites and can sometimes exclude nearly all other plants. The leaves are triangular to heart-shaped and coarsely toothed. The tall flower stalks have clusters of small white flowers with four petals. By early summer, most of the leaves have faded away, but the plants can still be recognized by the dead stalks with pale brown seedpods containing shiny black seeds.[11]

- Common reed, or phragmites, is a tall perennial grass that can grow to over three metres in height and form dense stands. The flowers form bushy purple or golden panicles in late July and August. Later in the season these panicles become grey and fluffy in appearance because of the hairs on the seeds. Phragmites can completely take over a marsh or damp meadow community.[12] There are both native and non-native subspecies of this plant in Ontario. The native strain grows in natural wetlands. Only the non-native variety is considered invasive, because it grows in disturbed wetlands and similar disturbed, open, moist sites.[13]

- Giant hogweed is also becoming more prevalent in southern, eastern, and central Ontario. It typically grows to heights of two to five metres and has a thick, dark reddish-purple stem. Even the leaves can grow to more than one metre in width. The hogweed's flowers are white and clustered together. Giant hogweed has the potential to spread into all types of open and semi-open habitats. The clear, watery sap of the hogweed contains toxins that can cause severe skin inflammation when the skin is exposed to sunlight.[14]

TALLGRASS PRAIRIES AND SAVANNAHS

Tallgrass prairies and savannahs are natural grasslands with a great diversity of grasses, wildflowers, and animal life. They are a globally imperiled ecosystem and one of the most endangered ecosystems in Canada. Only a few remnant sites remain. This type of habitat once covered large sections of southern and central Ontario from Windsor to as far east as Trent River near Rice Lake in Northumberland County. Tallgrass prairie typically includes native grasses such as big bluestem, little bluestem, Indian grass, and switch grass and wildflowers such as wild bergamot, prairie buttercup, wood lily, showy tick-trefoil, and butterfly milkweed. New Jersey tea and black oak are two woody plants that are also characteristic of this habitat. Native peoples camped in these prairie meadows, deliberately burning them to keep woody plants from becoming established. This allowed the prairie plants to thrive. When white settlers first arrived in the Rice Lake area, it is estimated that about 150 square kilometres of the land south of Rice Lake was tallgrass prairie. In fact, the native name for Rice Lake was "the lake of the Burning Plains," most likely referring to the fires that would have been set to kill woody plants encroaching on the prairie.

Located south of Rice Lake on the eastern edge of the Oak Ridges Moraine, Alderville First Nations (AFN) Black Oak Savanna/Tallgrass Prairie is the largest single remaining parcel of tallgrass prairie and savannah in east central Ontario. Through AFN's efforts at restoration by planting native prairie grasses and wildflowers using seeds collected on the site, the 48.5-hectare property is now the largest and best example of healthy tallgrass habitat in the Rice Lake Plains natural area.[15] The Goodrich Loomis Conservation Area, located north of Brighton, also contains numerous small prairie remnants and an oak savannah that is presently being restored. World Wildlife Fund Canada and the Ontario Ministry of Natural Resources have developed a recovery plan for Ontario's tallgrass communities and numerous restoration projects are now underway. These projects will also benefit a large number of increasingly threatened grassland birds like the Bobolink and rare butterflies like the mottled duskywing.

Colourful Wetlands

A skein of colour also brightens our wetlands this month. Among the taller shrubs and herbaceous plants, white flowers prevail as meadowsweet, tall meadow-rue, and common elderberry are all in bloom. Cattails, too, are still blooming early in the month. Pockets of open water are graced with bladderworts, yellow pond lilies, and fragrant white water lilies, the latter looking like so many china teacups. Along the edges of ponds and streams, the array of colours is completed by the blue spikes of pickerelweed and the white flowers of arrowhead. Just as the orange flowers of spotted jewelweed start blooming in mid-July, swamp milkweed, Joe-pye weed, and purple loosestrife add pinks and purples to the colour spectrum. According to the Ontario Ministry of Natural Resources, *Galerucella* beetles have been shown to control purple loosestrife, an invasive wetland plant. Beetles have been released in more than 400 locations across Ontario since 1993. The beetles have reduced the abundance of purple loosestrife in more than 80 percent of the release sites.[16]

Weather

July Highlights

All Month Long

- With hot, humid, and often thundery weather, July is prime lightning season. (See page 178)

- Local lakes usually reach their warmest temperatures. The average is about 23°C. This warm water sits on top of much colder water, almost creating two separate lakes. (See page 179)

- We still enjoy more than 15 hours of daylight for most of the month. By month's end, however, the sun is setting about 20 minutes earlier than on July 1.

- July is the warmest month of the year in central and eastern Ontario.

- The smells of July include the honey-sweet perfume of milkweed and American basswood blossoms. Leaves, too, contain a cornucopia of smells when rubbed. A few to try include wild bergamot (the peppery, citrus-like aroma of Earl Grey tea) and prickly-ash (lemon). The smell of smoke from a campfire tells us high summer has arrived, as well.

- Even in summer, temperatures on the Bruce Peninsula are among the most temperate in Canada. This is because both Georgian Bay and Lake Huron moderate the Peninsula's climate. Temperatures in the summer months (June–August) average 16.8°C daily.

- On hazy, humid summer days the sky appears pale blue, sometimes almost white. This is because the more water there is in the air, the more the blue wavelengths of light are scattered, leaving a paler looking sky. A similar phenomenon can occur in winter. (See January)

- So-called "heat lightning" is often seen. This is actually normal thunderstorm lightning but it is taking place too far away for its thunder to be heard. When a lightning bolt flashes, its light can be seen as far as 160 kilometres away, but the sound will carry for 25 kilometres at the most.

JULY WEATHER AVERAGES, 1971–2000[17]

City or Town	Daily Max. (°C)	Daily Min. (°C)	Rainfall (mm)	Snowfall (cm)	Precipitation (mm)
Owen Sound	24.5	14.9	72.6	0	72.6
Huntsville	25	13.8	84.2	0	84.2
Barrie	26	15	73.4	0	73.4
Haliburton	24.9	12.9	78	0	78
Peterborough	26.5	14.4	68.4	0	68.4
Kingston	24.8	15.7	58.8	0	58.8
Ottawa	26.4	15.5	88.9	0	88.9

STORMY WEATHER

With the month of July comes weather that is often hot and muggy. This is because almost half of our summer air masses originate over the Gulf of Mexico. The heat and humidity often lead to severe thunderstorms on July afternoons. As the air warms up over the course of the day, evaporation increases dramatically. The warm, moist air rises and forms white, fluffy cumulus clouds. However, if the moisture in the clouds increases too much, they will transform themselves into dark, flat-topped cumulonimbus, or thunderstorm clouds. These clouds take on an anvil shape when they reach a height at which they can rise no more. They then spread out sideways as there is nowhere else to go. Clouds stop rising when the density of the rising warm air becomes the same as the static air in which they are rising. Air expands as it rises because pressure drops. In doing so, it loses heat, sees its temperature fall, and the moisture condenses. The anvil part of the cloud is often at such a height that it is formed of ice crystals instead of water droplets.

The tops of the clouds build up a positive electric charge while the bottoms develop a negative charge. When the buildup of electrical charges becomes great enough, charges between parts of the cloud or between the cloud and the earth — which normally has a negative charge — are released as lightning. The huge amount of heat from lightning creates shock waves in the air that we call thunder. By remembering that thunder travels one kilometre in three seconds, the time between when you see the lightning and when you hear the bang will tell you how far away the storm is. However, according to Environment Canada, anytime you can hear thunder, you are within striking distance of lightning and should take shelter immediately, preferably in a house or automobile. This is because many lightning strikes occur at considerable distances from a storm system, even when clear skies are visible. Stay away from tall objects such as trees and poles. In addition to lightning and heavy rain, thunderstorms can also be accompanied by hailstones and high winds, and can cause locally extensive damage.[18]

Two Lakes in One

Central and eastern Ontario lakes usually reach their warmest temperatures — normally about 23°C — during the month of July. About five metres below the warm surface water, however, there is a much colder body of water that remains at 4°C all summer long. Anyone who has made a deep dive into a summer lake will attest to this fact. Because of the difference in water density, the two temperature zones do not mix. A summer lake is therefore best thought of as two separate lakes — a warm one on top and a cold one underneath. The area separating the two is referred to as the thermocline.

Oxygen levels in the cold water zone often become quite low in summer. First and foremost, oxygen from the air has no way of entering these cold, deep waters. As explained above, there is no mixing in summer between the oxygen-rich warm water zone and the cold water below. Second, because there is no light penetration into deep water, there is no plant growth, and consequently no oxygen is produced through photosynthesis. Finally, when algae from the surface waters die, they sink into the cold water zone. Valuable oxygen resources are used up in the process of decomposition. Low oxygen levels have a major impact on fish and invertebrates living in the cold water zone. In order to find sufficient oxygen, they are sometimes forced upward to the thermocline and can end up all crowded together in a narrow band of water. Lacking the necessary adaptations, some species cannot survive if they are forced to move into the warmer water above the thermocline.

APPROXIMATE JULY SUNRISE AND SUNSET TIMES (DST)[19]

(Note: Twilight starts about 30 minutes before sunrise and continues about 30 minutes after sunset.)

Location	Date	Sunrise (a.m.)	Sunset (p.m.)
Grey-Bruce	July 1	5:42	9:13
	July 15	5:52	9:07
	July 31	5:59	9:01
Muskoka/Haliburton	July 1	5:38	9:07
	July 15	5:48	9:01
	July 31	6:04	8:46
Kawartha Lakes	July 1	5:33	9:01
	July 15	5:43	8:56
	July 31	5:59	8:41
Kingston/Ottawa	July 1	5:26	8:54
	July 15	5:36	8:48
	July 31	5:52	8:33

NIGHT SKY

JULY HIGHLIGHTS

All Month Long

- Major constellations and stars visible (July 15 — 10:00 p.m. DST)

  Northwest: Ursa Major dominates the northwest sky with Ursa Minor (with *Polaris*) to its right and Bootes (with *Arcturus*) to its left; Leo (with *Regulus*) at horizon.

  Northeast: Summer Triangle made up of the bright stars of *Vega* (in the constellation Lyra), *Deneb* (in the constellation Cygnus), and *Altair* (in the constellation Aquila). Also look for Cassiopeia and the Milky Way; Great Square of Pegasus (low in northeast).

  Southeast: Sagittarius (near horizon) and above it, the brightest part of the Milky Way.

  Southwest: Bootes (with *Arcturus*) in mid-sky; Virgo (with *Spica*) to lower right.

- The Milky Way and Summer Triangle dominate the night sky. (See page 181)

- The Algonquian name for the full moon of July is the Buck Moon.

- Pegasus, the signature constellation of fall, becomes visible along the northeastern horizon in the late evening. Its arrival reminds us to enjoy summer now because it won't last!

- Generally warm and pleasant weather makes for comfortable star-gazing. However, you'll have to wait until about 10:00 p.m. for serious observing.

- Being opposite the high-riding summer sun, the summer moon travels low in the southern sky. This means that summer moon shadows are much longer than those of winter. The low moon also makes for romantic, long moonbeams over the water.

THE SUMMER TRIANGLE

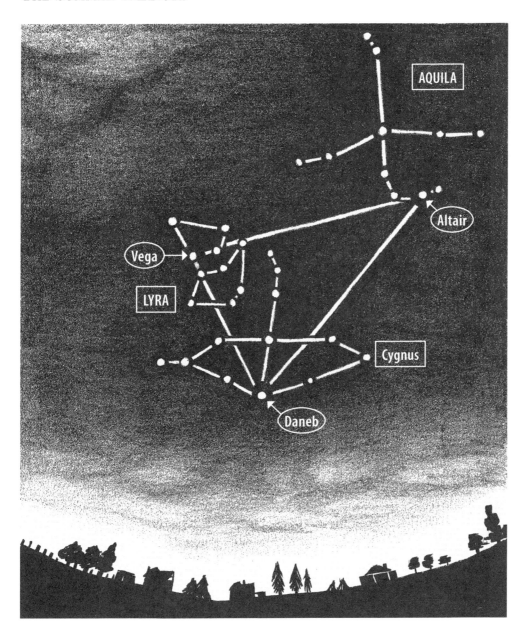

Knowing the constellations and asterisms like the Summer Triangle adds something magical to the enjoyment of summer nights.

Jean-Paul Efford

181

One of the main features of the summer sky is an asterism known as the Summer Triangle. It is made up of the brightest or alpha stars of three separate summer constellations. By joining these stars together, a giant triangle is formed. A star-rich swath of the Milky Way runs right through the Summer Triangle and continues to the horizons in both directions. To find the Summer Triangle, use a sky chart along with the Big Dipper as your guides. All summer long, the Big Dipper is suspended high in the northwest. The two stars that form the end of the Dipper's bowl closest to the handle point almost directly to Deneb, one of the three stars of the Triangle. The other two stars in the Triangle are Vega and Altair. Deneb marks the tail of Cygnus, the Swan, an easily recognizable constellation. The central stars of Cygnus are also known as the Northern Cross, a shape they closely resemble. Altair forms the head of Aquila, the Eagle, which looks like a giant bird soaring across the heavens on outspread wings and almost on a collision course with Cygnus. Although Vega is located in a rather small, unremarkable constellation (Lyra), it is the brightest star of the Summer Triangle and the second brightest star in the summer sky. Only Arcturus, which is now high overhead, is brighter.

AUGUST

Summer Becoming Fall

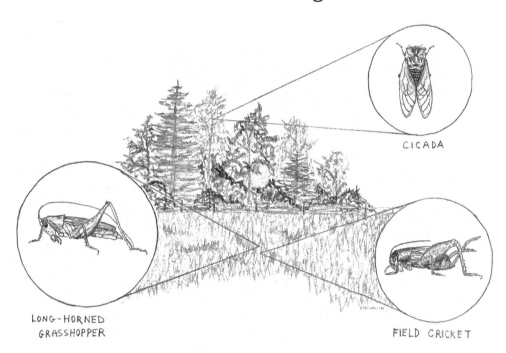

CICADA

LONG-HORNED
GRASSHOPPER

FIELD CRICKET

Three key musicians in the August soundscape.
Kim Caldwell

The real voyage of discovery consists not in seeking new landscapes but in having new eyes.

— Marcel Proust

THE FRANTIC PLANT GROWTH AND ANIMAL ACTIVITY OF SPRING AND EARLY SUMMER have now been replaced by a languid atmosphere of maturity and calm. But, despite weather which is often hot and sultry, August is very much "summer-becoming-fall." Bird migration is well underway, the first leaves are starting to change colour, and roadsides are being transformed by a yellow surf of goldenrod. August days can seem strangely silent, since songbirds have finished nesting and no longer need to declare their territories. The silence in the trees, however, has been replaced by a musical frenzy in meadows and marshes as countless crickets and grasshoppers broadcast their sexual desire through buzzes and clicks of every description.

Our other senses this month are piqued by the spicy fragrance of wild bergamot, by the delicious taste of fresh corn and tomatoes, and by the calming sight of misty dawns. For many of us, there is also the all-too-familiar irritation in our eyes, nose, and throat as ragweed pollen triggers another hay fever season.

Thoreau observed "how early in the year it begins to be late." Already, signs of fall are everywhere. On clear, cool evenings we hear the contact calls of migrant songbirds as they stream overhead against the backdrop of the Milky Way, while snowy tree crickets call in perfect unison in the background. With late August comes the anticipation of bright, cool September weekends and the riot of colour that is just around the corner. In a cultural sense, August is much more the end of the year than is December; because, with Labour Day, our lives begin anew with everything from a new school year to the reconvening of myriad community activities.

BIRDS

AUGUST HIGHLIGHTS

All Month Long

- Many songbird species are moulting and feeding heavily in preparation for migration. Typical of these are Bobolinks, which have mostly gathered into flocks and taken refuge in marshes. (See page 185)

- Shorebird numbers swell to at least ten species. Some shorebirds show up in wetlands in late summer, when low water levels expose the muddy, invertebrate-rich bottom. (See page 187)

- Cedar Waxwings are conspicuous August birds. Watch for them perched on the branches of dead trees as they sally out to catch insects on the wing. (See page 186)

- Swallows continue to flock up on wires. Most will have departed by month's end.

- On our larger lakes, it is not uncommon to see a half-dozen or more Common Loons swimming together. They sometimes dive and resurface in unison, rear up and flash their wings and appear to talk to one another in chuckles and croonings. It is thought that they may be non-breeding individuals or perhaps birds that lost their nests early in the season and did not re-nest. The exact purpose of these gatherings is not fully understood, but they certainly appear to have a social function.

- With its reproductive purpose now completed, birdsong has faded to only a shadow of what it was in the spring. Other than the sporadic singing of a handful of species such as Eastern Wood-pewees, Red-eyed Vireos, Northern Cardinals, Song Sparrows, and Mourning Doves, most song has ceased.

- Common, however, are the high-pitched "lisping" calls of Cedar Waxwings, the harsh *jaaaay* screams of Blue Jays, the cawing of crows, and the *po-ta-to-chip* flight call of the American Goldfinch. Birds calls — as opposed to songs — have many purposes, such as signalling danger and keeping the flock together.

- Ospreys are very vocal. Listen for their short, chirping whistles, especially around the nest or as they soar overhead.

Mid-August

- Songbird migration is in full swing by mid-month, with numerous warblers and vireos slowly making their way southward. (See page 187)

- Listen for the faint contact calls of migrating birds as they pass overhead at night. Because each different type of bird has a unique call note, it is sometimes possible — albeit with lots of practice! — to identify some of the species flying over. The silhouettes of nocturnal migrants can sometimes be seen passing in front of a bright moon.

- Large flocks of Ring-billed Gulls as well as Killdeers are often seen feeding in recently ploughed fields during the late summer and fall. Other species of shorebirds sometimes turn up as well.

PREPARING FOR MIGRATION

In the late summer, songbirds prepare for migration by undergoing important physiological changes. As the days shorten, photoreceptors in the brain respond to the decreasing amount of daylight by triggering hormonal changes. These changes cause the birds to eat voraciously and to become increasingly restless. Many songbirds abandon an insect diet and start to gorge themselves on fruit, which is more easily converted to energy-rich fat to fuel the flight southward. Many species also begin a complete moult of their feathers and return to the "basic plumage" that will take them through the fall and winter. Moulting is necessary because feathers become worn and therefore less effective in supporting flight and keeping the bird warm. Moulting can take from five to 12 weeks to complete in most songbirds. In late winter or spring, these birds will moult once again into a brighter "alternate plumage" for the breeding season. Moulting birds can sometimes have a strange combination of both plumages and therefore look quite different from anything in a field guide. Moulting is a dangerous and energetically expensive time for birds and many species become difficult to find.

Ducks are somewhat unique when it comes to moulting. They attain their duller basic plumage right after breeding and keep it for only a few months. The males usually acquire their bright alternate plumage by late fall and will often begin courtship on their wintering sites. It is important to note, however, that the terminology that should be used to describe moulting in ducks is still being debated. Some recent studies suggest that what we call basic plumage in ducks may actually be their alternate plumage, and vice versa.

WAXWINGS

Each month of the year has its characteristic birds that take centre stage. In August, one of the most visible and often-heard species is the Cedar Waxwing. Flocks of these black-masked berry lovers are common along roadsides, watercourses, and even in suburban backyards, especially if there is fruit available for them to feed on. Their behaviour, too, is very predictable. If you see a small group of birds perched high in a dead tree at this time of year, the chances are very good that they are waxwings. Listen for a very high-pitched, sibilant *seee-seee-seee* call. It is especially noticeable when the birds take wing. Like the drone of the cicada, the call of the waxwing is a characteristic sound of hot, languid August days.

If you are at the cottage, out paddling, or even driving alongside a body of water, watch for waxwings making aerial sallies from exposed perches to catch species such as dragonflies and mayflies emerging from the water. This behaviour is referred to as hawking. Flying ants are also a popular prey item, especially when the ants form mating swarms in early September. In the fall, waxwings also rely on wild grapes, ornamental fruits such as crabapple and mountain ash, and the "berries" of red cedar trees — hence the cedar part of their name. These are not true berries, however. After the female cones of the red cedar — actually a type of juniper — are pollinated in the spring, the scales swell up and fuse, thereby forming a small, soft, fleshy cone that superficially looks like a berry. Cedar Waxwings are also different from most other birds in that they are very late nesters. In fact, some individuals do not begin egg-laying until early August. Lateness in breeding has probably evolved as a result of the species' dietary reliance on fruits that don't ripen until summer.

SHOREBIRD AND SONGBIRD WATCHING

Birding can be excellent in August, especially in the second half of the month when both songbirds and shorebirds are migrating in large numbers. Common shorebird species to be expected include Lesser and Greater Yellowlegs, Least, Semipalmated, Pectoral, and Solitary Sandpipers, and Semipalmated Plover. Shorebirds are especially interesting because they allow the observer to get quite close and to take long, leisurely looks. There is also a special aesthetic in their tightly knit flocks that twist and turn on a dime, their restlessness, and the vast distances they travel. These are birds that represent the ends of the earth. When you are watching a solitary Sanderling on an August day, you can't help but feel moved by a 50-gram bundle of feathers that was born mere weeks before on Baffin Island and will fly all the way to southern Argentina without the aid of adult birds to show the way.

Finding migrant shorebirds in much of central and eastern Ontario, however, can be a challenge. Other than in the spring, when yellowlegs often turn up in flooded fields, we do not have a lot of quality shorebird habitat in this region. If you are lucky, you may find small numbers of summer and fall migrants at wetlands when low water levels expose the muddy, invertebrate-rich bottom. However, to be relatively certain of seeing a good mix of species, a trip to a local sewage lagoon or to the shores of the Great Lakes or Ottawa River will probably be necessary.

By mid-August, songbird migration is very much in evidence, as well. Unlike the spring, fall songbird migration is a much more drawn out process, with many species moving through the region over a period of a month or more. For example, the Red-eyed Vireo is a common migrant from the middle of August until the end of September. Most warbler species, too, will continue to migrate through until at least the middle of September. Migration is initiated by the arrival of cool, damp weather with northerly winds. Migrants often turn up along the margins of lakes, in the trees and shrubs along cottage roads, along shrubby fencerows, and even in city backyards. However, much of late summer and fall songbird migration tends to be a quiet and somewhat secretive phenomenon. The observer therefore needs to slow down, to watch carefully for any movement, and especially to listen. Although the birds are not singing in the fall, they are making contact calls. Migrants also tend to be in mixed flocks, which often include very vocal chickadees. A rule of thumb of fall birding is to stop and take a close look any time you hear chickadees calling because there are usually other species with them. Be sure to try pishing in order to draw the birds in for close-up looks. Keep the pishing up for at least a minute or two. You will marvel as chickadees and then warblers of a half-dozen species or more sometimes come within a couple of metres of you.

The Semipalmated Sandpiper breeds on the open tundra of the Canadian Arctic.

Karl Egressy

MAMMALS

AUGUST HIGHLIGHTS

All Month Long

- Wolves and coyotes are quite vocal this month. The young in particular will sometimes even respond to human imitations of their howls. Central and eastern Ontario wolves have now been identified as a separate species, namely the eastern wolf. (See page 189)

- A large variety of mammals gorge themselves on late summer fruit and nuts. Bears are especially fond of American beech nuts and sometimes leave large piles of broken branches called "bears' nests" high in the crotches of trees where they have been feeding. Coyotes, on the other hand, are very fond of apples.

- Gray squirrels may give birth to a second litter this month.

- Eastern chipmunks continue to be extremely vocal. They make an incessant, bird-like "chuck-chuck-chuck" sound, the purpose of which may be to advertize ownership of territory.

Late August

- Little brown bats begin to congregate in large numbers at mating and hibernation sites such as caves and old mine shafts.

COYOTES AND WOLVES IN CENTRAL AND EASTERN ONTARIO

"If all the creatures in the world were to die, the coyote would be the last one left." This is how one Native American legend describes the supreme ability to survive of *Canis latrans*, better known as the western coyote. The eastern form of this amazingly adaptable canine is common in central and eastern Ontario, although much of the time its presence still goes undetected. This is too bad because they are beautiful animals. The eastern coyote is about the size of a small, lean German Sheppard dog. They are grizzled grey with a whitish throat and belly and rusty ears and legs. The tail is long and bushy. Coyotes keep their tail down while running, whereas wolves tend to hold theirs straight out. On average, eastern coyotes weigh about 16 kilograms which is four kilograms more than western coyotes. The reason for their larger size in eastern Canada stems from their hybrid origin with eastern wolves.

Through genetic analysis done by Brad White and Paul Wilson of Trent University, it has now been established that the wolves of central and eastern Ontario, including Algonquin Park, are not a small race of the gray wolf, as previously believed. Rather, they are a totally separate species, similar to the red wolf of the southern United States.[1] They have now been given their own name, that of eastern wolf (*Canis lycaon*). According to White and Wilson, the red wolf, the eastern wolf, and the coyote evolved from a single North American ancestor.

The eastern wolf is smaller than the gray wolf and has a grey-reddish coat with black hairs covering the back and sides of the thorax. The back of the ears is reddish.

The eastern coyote owes its large size to interbreeding with wolves.
Tyler Wheeldon

Western coyotes are closely related to the eastern wolf; consequently, the two species readily hybridized when western coyotes expanded into southwestern Ontario in the early part of the 20th century, a male wolf mating with a female coyote. The presence of wolf genes explains why eastern coyotes are bigger and darker than their western cousins. In fact, all wolf and coyote-like animals in central and eastern Ontario contain, to varying degrees, both coyote and eastern wolf genetic material. Depending on the habitat, one or the other is dominant. In more open, agricultural areas, the animals contain more coyote genes. In forested, northern areas, wolf genes are more dominant and the animals are slightly larger and heavier. Even the wolves of Algonquin Park show some evidence of prior hybridization with coyotes.

Coyotes and wolves play an important role in the ecosystem. By killing raccoons, for example, coyotes indirectly benefit songbirds and turtles whose eggs are often preyed upon by this species. By preying on deer, wolves help the many other species that suffer from habitat degradation caused by exploding deer numbers.

AMPHIBIANS AND REPTILES

AUGUST HIGHLIGHTS

All Month Long

- Leopard frogs wander en masse from their wetland habitat to invade nearby fields to feed on the bounty of late-summer insects.

- Wood frogs and spring peepers are fairly easy to find in the lower, damper areas of woodlands in late summer and early fall.

- American toads become more common on our lawns and in our gardens. Young leopard frogs and green frogs often show up as well.

- Green frogs and gray tree frogs may continue to call sporadically.

- Massasauga rattlesnakes give birth to seven to ten live young in late July or August. Young snakes are sometimes seen in wetlands in the summer.

Late August

- Since the eggs of all of our turtle species hatch from late August to early October, it is not uncommon to come across a baby turtle. Watch for them along rail-trails and roads that pass by wetlands. However, we usually don't see young painted turtles in the fall because most actually stay in the ground until the following spring after hatching out of the egg.

FISH

AUGUST HIGHLIGHTS

All Month Long

- Lake, Brown, and Brook Trout feed heavily in preparation for fall spawning.

- Many species of fish continue to seek out deeper, colder water. These include Walleye, Muskellunge, Smallmouth Bass, Lake Trout, and Brook Trout. The Walleye's large eyes allow it to see well, even in the low light conditions of deep water. In the deeper Shield lakes of northern central and eastern Ontario, Lake Trout now frequent the cold waters below the thermocline (the transition layer between the warmer surface water and the much colder deep water) where they find the temperatures of 10°C and colder that they prefer.

- Anglers often concentrate their efforts on bass this month.

INSECTS AND OTHER INVERTEBRATES

AUGUST HIGHLIGHTS

All Month Long

The dog-day cicada, *Tibicen canicularis,* can be identified by the greenish markings on the mostly black body.

Cody Hough, Wikimedia

- The sounds of countless crickets, grasshoppers, and cicadas fill the natural world. How they produce their "songs" is fascinating. (See page 193)

- A large percentage of the insect music we here this month comes courtesy of crickets and katydids. For example, the snowy tree cricket sounds like a gentle-voiced spring peeper. Its beautiful rhythmic pulsations actually provide a good estimate of air temperature. (See page 194)

- Wasps seem to be everywhere, especially where food is present. (See page 194)

- Ants are abundant this month, even on the granite rocks of the Canadian Shield. Most of these are formicine ants and are usually either black or black and red.

- Although fewer species of butterflies are seen in mid to late summer, monarch numbers increase in August and by month's end sulphurs are very plentiful.

- Milkweeds attract an interesting variety of invertebrate life. Watch for the tuft-covered milkweed tussock moth caterpillar, the red and black large milkweed bug, the eastern milkweed longhorn beetle, and especially the yellow, black, and white monarch butterfly caterpillar. These caterpillars are easy to rear in captivity and provide adults and children alike with a front row seat to the marvel of insect metamorphosis.

- Grasshoppers, also known as locusts (family Acrididae), are abundant. Walk along any dusty road in August and you can't help but see and hear Carolina locusts (*Dissosteira carolina*). They are conspicuous because of the crackling noise they produce in flight and the eye-catching pale border on the dark hindwings. Also watch for the red-legged grasshopper (*Melanoplus femurrubrum*), a generally yellow grasshopper with red legs. It does not produce any audible sound, however.

- The electric, buzzing sound of the cicada seems to drain the energy right out of you. The most commonly heard species of cicada in central and eastern Ontario is the dog-day cicada (*Tibicen canicularis*), probably named because its emergence and calling often coincide with the hot and muggy "dog days" of summer.

- Watch for Catocala, or underwing moths, named for the bright colours of the underwings. However, when the underwings are hidden, all that is visible is the mottled brown or grey of the forewings, which look very similar to bark.

- Cottagers sometimes find large, mysterious jelly-like "blobs" attached to the dock or aquatic plants. They are formed by colonies of Bryozoa, a freshwater invertebrate. Looking somewhat like an egg mass, the clumps are clear, dense, and have distinct, repetitive patterns and markings on the outside. Bryozoa are like a freshwater coral in that the mass they form is actually a colony of thousands of zooids — roughly analogous to polyps in corals. Each tiny zooid has whorls of ciliated feeding tentacles that sway back in forth to catch plankton in the water.

Late August
- Small meadowhawk dragonflies of the genus *Sympetrum* are a common sight. In most species, the males are red and the females are yellow.

THE INSECT CHORUS HITS ITS STRIDE

From the middle of summer until the first frosts, the natural soundscape is dominated by the incessant calls of crickets, katydids (long-horned grasshoppers), and grasshoppers (locusts or short-horned grasshoppers), all of which belong to the order Orthoptera. Cicadas (order Homoptera) add their imposing voices to the din, as well. As with birds and amphibians, only male insects call. The express purpose of their calls is to attract a female for the purpose of mating. As a general rule, we hear crickets and katydids during both day and night, while cicadas are vocal only during the day. Most grasshoppers are not as vocal as their Orthopteran cousins.

Crickets, katydids, and grasshoppers use stridulation to produce their song. In much the same manner as a violin string being scraped by a bow, these insects rub one part of their body on another. Katydids and crickets elevate their forewings and move them back and forth rapidly. The base of one wing has a hardened edge, or "scraper," while the base of the other has a toothed ridge, or "file." Each time the scraper hits a tooth, a click is produced. This causes the wings themselves to actually vibrate and produce the "song." Because the rubbing occurs so fast, the individual clicks blend together and sound like chirps or trills. In the case of grasshoppers, only one group, the slant-faced grasshoppers, actually stridulate. They do so by rubbing one of the hind legs over a projecting vein, or scraper, on the forewing. Each hind leg has a row of about 80 fine spines which vibrate like the teeth of a comb. Sounds produced in this manner are typically soft and muffled. Another group, the band-winged grasshoppers, sometimes snap their wings together in flight to produce crackling sounds.

Cicadas employ a totally different technique to attract females. Males have a pair of special sound-producing organs known as "tymbals." They are located on the sides of the abdomen, just behind where the hindwings are attached. By contracting muscles, ribs in the tymbals bend suddenly and produce a vibrating click. The clicking noises are amplified by a large air sac in the abdomen. Cicada song far surpasses that of most Orthopterans in terms of volume. When a female finds a male that interests her, she gives a flick of her wings which stimulates the male to come closer.

To listen to crickets, katydids, and grasshoppers, the best locales are marshes and meadows, particularly on a sunny afternoon or warm evening. At night, the shrubby edges of woodlands are also excellent. Cicadas are common just about everywhere there are trees. Whereas identifying many Orthopteran and Homopteran species by sight is next to impossible, distinguishing at least some of them by their songs is usually relatively easy. *The Songs of Insects*, by Lang Elliot and Wil Hershberger, is an excellent book for identifying insects by their call. It includes a CD of insect recordings. (See Bibliography)

CRICKETS AND KATYDIDS

Around houses, crickets are often the most common insect songsters. The best-known members of this group are the field crickets (genus *Gryllus*), which are the robust black crickets with round heads. Most of the cricket song we hear, however, comes from a smaller group known as ground crickets. Less than one centimetre in length and looking like "baby" field crickets, you often see them scurrying to get out of the way as you walk across a lawn or meadow. This family of crickets creates a non-stop wall of sound both day and night. Listen for a continuous trill that differs in rate and sound quality, depending on the species. A rapid but soft series of very high "tikitikitiki…" notes is typical.

Tree crickets (genus *Oecanthus*) are another very vocal group. Unlike field and ground crickets, tree crickets are delicate, translucent green insects with lacy wings. Most sing with a prolonged trill, reminiscent of the American toad. One species of particular interest is the snowy tree cricket (*Oecanthus fultoni*), which delivers a steady stream of melodious chirps at dusk on warm evenings. Its rhythmic "treet, treet, treet" is often described as sounding like a rich, gentle-voiced spring peeper. It is definitely one of the most beautiful sounds of late summer. Often several males will sing in perfect unison. This species is also known as the thermometer cricket because it is possible to calculate the air temperature by the frequency of its calls. By counting the number of chirps in eight seconds and adding five, the temperature in degrees Celsius can be estimated quite accurately. Roadsides and other brushy areas adjacent to deciduous woodlots are often good places to listen for them. They can sometimes be heard in suburban areas, as well.

Katydids (long-horned grasshoppers) belong to the Tettigoniidae family. Because they are a less active group and their green coloration provides excellent camouflage, they are not well known to most people. They all have long, filamentous antennae. Central and eastern Ontario is home to a number of different kinds of katydids. One group, known as bush katydids (genus *Scudderia*), typically produce short, electric "teeth of a comb" sounds, reminiscent of a chorus frog. The true katydids (Genus *Pterophylla*), however, are not found in central or eastern Ontario. The common true katydid (*Pterophylla camellifolia*) is the insect that most people associate with the name "katydid." Its raucous "katy-did … katy-did" is well known throughout most of the eastern United States and extreme southern Ontario. With climate change, however, common true katydids may extend their range northward.

WASPS, YELLOWJACKETS, AND HORNETS

As pleasant as a late summer picnic might be, there always seems to be a handful of unwanted guests. Like corn-on-the-cob, ripe tomatoes, and blueberry pie, hornets and yellowjackets are often part of any late summer outside meal. They would be little more than an annoyance if it wasn't for the fact that they can deliver a painful sting. Hornets and yellowjackets are members of a family of insects known as the Vespidae, which, in turn, is part of the much larger order Hymenoptera, or bees, wasps, and ants. Vespidae are commonly referred to as wasps, as well. They can be recognized by the way they hold their wings slightly out to each side. One of the most commonly seen species is the eastern yellowjacket (*Vespula malculifrons*). It is easily identified by the triangular black marking nearest the thorax. It has a narrow black stem or neck which extends to the upper edge of the abdomen. Yellowjackets usually nest in the ground, often in an abandoned animal burrow. They are particularly abundant after a dry, warm summer because these conditions allow the nests to flourish. At summer's end, there is a frantic search for food to feed the thousands of larvae still in the nest. Caterpillars are the larval food of preference but yellowjackets will also turn to human foods as a source of protein for the colony. Adults also need sugar in order to fuel the energy requirements of their own bodies. Some of their favourite sources of sugar include flower nectar (especially goldenrod), ripe fruit, and aphid honeydew (usually gleaned from tree leaves). However, as we know all too well, wasps are also attracted to the same sweet drinks as humans.

Large black-and-white Vespids called bald-faced hornets (*Dolichovespula maculata*) are also very common. They have a mostly black body with yellowish-white markings on the side and face. Hornets draw particular attention because of their habit of building globular, paper nests in trees. A large colony can harbour up to 600 individual hornets by September. This species needs to be treated with respect. Only mated queen Vespidae survive the winter. The rest of the colony succumbs to the first hard frosts of the fall.

Like all living creatures, wasps play an important ecological role. First of all, many species are important pollinators. A group known as pollen wasps have actually substituted pollen for insect prey and, like bees, feed it to their larvae. You can easily find wasps feeding on the pollen and nectar of goldenrod flowers. In the feeding process, the wasps are inadvertently transferring pollen from one flower to another.

PLANTS AND FUNGI

AUGUST HIGHLIGHTS

All Month Long
- Red maples along the edge of lakes and wetlands are the first trees to show splashes of fall colour. Virginia creeper and staghorn sumac also provide a hint of the colour to come. (See page 197)

- There is often a very noticeable algal bloom in August or September. The surface water may turn almost completely green. (See page 198)

- A profusion of ripe wild fruits can be found on various trees, shrubs, and vines this month. They will help to fuel much of the fall bird migration. (See page 199)

- Although shady woodlands are now devoid of flowers, a large variety of shade-tolerant ferns has taken their place. They are fun to try to identify. Among the more common species to expect are marginal wood fern, ostrich fern, maidenhair fern, sensitive fern, Christmas fern, lady fern, and interrupted fern.

- The uncommon but beautiful cardinal flower adds new reds to wetland edges. The cardinal flower is a popular nectar source for Ruby-throated Hummingbirds.

- The bladderworts, a group of aquatic plant with small pink or yellow flowers, are now in bloom in local wetlands. Their carrot-like leaves can be seen on or just below the surface of the water. Green sacks called bladders are scattered among the leaves' many branches.

- Many leaves now have a dusty, tattered look. The leaves of Norway, silver, and Freeman's maple (a hybrid of silver and red maple) may also develop black spots as a result of a fungal disease called Tar Spot. The overall health of the tree does not appear to be affected. You may, however, want to consider replacing your Norway maple (an exotic species that can invade natural areas and shade out native plants) with a native tree.

- Fireweed seed pods open and their seeds fill the air, carried on silky plumes of white hairs.

- Pale corydalis blooms in pockets of soil on the granite rocks of the Canadian Shield all month long. The tube-shaped flowers are pink with yellow tips.

- Lower lake and river levels in late summer allow a suite of specialized plant species — including rare "Atlantic Coastal Plain" (ACP) plants such as Virginia meadow beauty — to germinate and thrive on the exposed sand and mud shores. However, the stabilization of lake water levels through the use of dams is a threat to those species that require water level fluctuation. Southeastern Georgian Bay is a well-known area for ACP species.[2]

Early August

- Queen Anne's lace continues to dominate roadsides but goldenrod numbers are increasing. The deep lavender of thistles, the bright yellow of smooth hawk's beard, and the pale pink of bouncing bet are also common sights along roads and trails.

- Evening primrose blooms in dry, open areas such as along roadsides. Unlike most other plants, the yellow flowers open their petals only in the evening, hence the plant's name. The flowers also release a strong fragrance that attracts moths, the only available nighttime pollinators.

Mid-August

- As long as there is sufficient rain and humidity, mid-August through most of September is usually the peak season for mushroom hunting. (See pages 200 and 201)

- Ragweed is in full bloom and its pollen has hay fever sufferers cursing with every sneeze. Goldenrod, which relies on insects to spread its sticky, heavy pollen, is not the culprit. The small green flowers of the ragweed, however, rely strictly on the wind to spread the ultra light, spike-covered pollen grains. A single ragweed plant can produce a billion grains of pollen! Density is highest during the morning hours. Research done by the U.S. Department of Agriculture has shown that over the past four or five decades the higher CO_2 levels associated with global warming may have doubled the amount of pollen that ragweed is producing.[3]

Late August

- Goldenrods reach peak bloom at month's end and take over as the main roadside and field flowers.

- Sugar maple and white ash keys ripen and begin to fall.

- The purple-black berries of the European buckthorn are ripe. This exotic invasive has dark green, glossy leaves and is by far the most common shrub in urban and agricultural areas. Native species have difficulty competing with buckthorn, however, and it sometimes forms near monocultures. Biodiversity suffers.

"FIRST BLOOM" CALENDAR FOR SELECTED TREES, SHRUBS, AND HERBACEOUS PLANTS

Early August: Turtlehead, great lobelia, virgin's bower, wild cucumber, small-flowered agalinis (gerardia), white snakeroot, woodland sunflower, large-leaved aster, flat-topped white aster, grass-leaved goldenrod.

Mid-August: Ragweed, purple-stemmed aster, Canada goldenrod.

Late August: Bur-marigold, bottle gentian, fringed gentian, New England aster, calico aster, heath aster, panicled aster, grass-of-parnassus, ladies' tresses orchids.

A HINT OF THE COLOUR TO COME

Along sunny roadsides and rail-trails, a hint of the fall colour to come is already present by late August. Virginia creeper is starting to offer up splashes of red, as are some of the red maples growing beside lakes and wetlands. Numerous choke and pin cherries have already turned orange, along with some of the leaves on staghorn sumac. Yellows abound in August, too, thanks mostly to the abundance of goldenrod. However, there is already a serving of yellow in the leaves, much of it courtesy of some precocious elms and balsam poplars. Some of the dogwoods have also jumped the gun by providing an early offering of purples and burgundies. Finally, one can't help but notice the appearance of orange and yellow leaves on a number of sugar maples in late summer. At this early date, colour change in this species is usually an indication of a severely stressed tree.

Blue-green Algae Blooms

Algae are tiny aquatic plants that contain chlorophyll for the purpose of manufacturing food. They are usually green in colour. Algae exist in many forms, from microscopic single cells to mass aggregates as well as forms that resemble higher plants. Ontario's waters are home to thousands of different species of these organisms, and they are critically important. Algae form the base of the food chain by converting nutrients into organic matter. If there were no algae, there would be no fish. Also, with one important exception, most algae do not produce substances that are toxic to humans. The exception is blue-green algae.

Blue-green algae, or cyanobacteria, are actually not algae at all. They belong to an ancient group of organisms that are most closely related to bacteria. However, like green plants, they rely on sunlight for energy through the process of photosynthesis. In fact, blue-green algae are given credit for the origin of plants. The chloroplasts that plants use to make food through photosynthesis are actually cyanobacteria living within the plant's cells. Being bacteria, blue-green algae are quite small and usually unicellular. The reason we can see them is because they often grow in large colonies. They are naturally present in all aquatic ecosystems, including lakes, rivers, and streams.

Like many of the true algae, blue-green algae can become extremely abundant in warm, shallow surface water that receives a high amount of sunlight and increased inputs of nutrients, especially phosphorus and nitrogen. When this sort of super-charged algal growth occurs, it is referred to as a "bloom." Blooms discolour the water and often produce a floating scum on the surface. At times, the water takes on an almost paint-like appearance and consistency. Thick pea soup immediately comes to mind. The wind will often blow the algae into bays, making it all the thicker. Most large blooms in central and eastern Ontario occur in late summer or fall but can also happen earlier in a dry, hot year. And, with climate change, drier, hotter summers are to be expected. Species of the genus *Microcystis* — a potentially toxic group of blue-greens — are commonly involved at this time of year. Some toxins produce allergic reactions such as rashes and eye irritation in sensitive people who come into direct contact with the toxins when swimming or showering. Other toxins have been found to affect the nervous system. This has been seen in animals that have ingested large amounts of water contaminated with blue-green algae. It's important to keep pets and livestock away from water if a bloom has occurred. Phosphorus, and to some degree nitrogen, are the limiting factors in algal growth. These nutrients are found in detergents, fertilizers, and in animal and human waste.

Blooms of true algae and those of blue-green algae are often confused. According to Agriculture Canada,[4] if you scoop a handful of the bloom with spread fingers and long, stringy masses are left dangling, it's most likely a true alga. If, after straining through your fingers, all that's left are bits and pieces sticking to your skin, it's probably a blue-green bloom.

Some Trees, Shrubs, Vines, and Wildflowers Bearing Fruit in August

(List organized by family)

Skunk currant

Chokeberry (*Aronia*)

Hawthorn

Apple

Pin cherry

Black cherry

Chokecherry

Blackberry

Red raspberry

Bristly sarsaparilla

Wintergreen

Purple-flowering raspberry

American mountain ash

Poison ivy

Staghorn sumac

Winterberry holly

Climbing bittersweet

European buckthorn

Virginia creeper

Riverbank grape

Alternate-leaved dogwood

Round-leaved dogwood

Bunchberry

Cranberry

Blueberry

Tartarian honeysuckle

Common elderberry

Maple-leaved viburnum

Hobblebush

Nannyberry

Downy arrow-wood

Highbush cranberry

Dogwood berries are high in fat and therefore provide important "fuel" to migrating birds.
Drew Monkman

199

MAGICAL MUSHROOMS

In much the same way as wildflowers epitomize a spring woodland, mushrooms signal the gradual approach of fall. When the weather becomes slightly cooler and we get an extended period of thick cloud cover and intermittent rain, mushrooms — the spore-producing, above-ground "fruits" of fungi — become abundant and appear in an incredible variety of shapes, colours, and sizes. The portion of the mushroom that you don't see is the hyphae. They are thin, string-like threads that form a mat in the ground or in decaying wood. The main mushroom groups include the gilled fungi or "true mushrooms," which have the typical flat or rounded cap; the coral fungi, which are highly branched; the polypores, which look like wooden shelves or brackets protruding from a tree trunk; the jelly fungi, which are rubbery to the touch and extremely colourful; and the puffballs which release their spores in a plume of "smoke" from a small opening in the top of the mushroom.

Although mushrooms have now been given their own kingdom (Fungi), we still often think of them as an element of our flora. In fact, they have very little in common with plants, being both biologically and ecologically completely different. Unlike plants, fungi do not contain chlorophyll and cannot make their own food. How they get their food is interesting. Many species have a symbiotic relationship with trees in which there is an exchange of services. The tree, by means of its roots, provides sugars for the fungus. In exchange, the fungus provides the tree with water and minerals. Some of the most common symbiotic varieties include *Amanitas*, *Boletes*, and *Russulas*. Most fungi are very limited in the number of tree species with which they can associate. The painted bolete (*Suillus spraguei*), for example, only lives in association with white pine. Another group of fungi known as saprophytes or decomposers digest dead organic matter, thereby allowing many of the nutritive elements in the organic matter to return to the soil. In this way, they play an essential role as recyclers in forest ecosystems. The shaggy mane mushroom makes a living in this manner as do the stinkhorns. Stinkhorns are distinctive because of their single erect stalk and the foul odour they produce. The odour attracts flies and other invertebrates, which inadvertently pick up and spread the stinkhorn's spores, present in a slimy substance located toward the tip of the mushroom. Finally, a third group of fungi are parasites. Parasites take nutrition from the tree without providing anything in exchange. Some grow on otherwise healthy trees and can live for many years at the tree's expense. Others only attack trees that are already dying. Polypores such as the tinder polypore (*Fomes fomentarius*) are parasites. After the tree dies, they are able to change lifestyles and become saprophytes.

Generally speaking, coniferous and mixed woodlands are richer in fungi than deciduous forests. However, stands of oak, beech, birch, and poplar are often an exception to this rule.

Familiar Summer and Early Fall Mushrooms
(List organized by genus)

Chicken of the woods
  (*Laetiporum sulphureus*)[E]

Turkey-tail
  (*Trametes versicolor*)

Artist's conk
  (*Ganoderma applanatum*)

Birch polypore
  (*Piptoporus betulinus*)

Purple-toothed polypore
  (*Trichaptum biforme*)

Meadow mushroom
  (*Agaricus campestris*)[E]

Destroying angel
  (*Amanita virosa*) — poisonous!

Yellow patches
  (*Amanita flavoconia*)

Fly agaric
  (*Amanita muscaria*) — poisonous!

Honey mushroom
  (*Armillaria mellea*)[E]

Shaggy mane
  (*Coprinus comatus*)[E]

Blewit
  (*Lepista nuda*)[E]

Dog stinkhorn
  (*Mutinus ravenelii*)

Giant puffball
  (*Calvatia gigantea*)[E]

Pear-shaped puffball
  (*Lycoperdon pyriforme*)[E]

Dung-loving bird's nest
  (*Cyathus stercoreus*)

Collared earthstar
  (*Geastrum triplex*)

Bog russula
  (*Russula paludosa*)[E]

Red mouth bolete
  (*Boletus subvelutipes*)

White pine bolete
  (*Suillus americanus*)[E]

Delicious lactarius
  (*Lactarius deliciosus*)[E]

E: *edible species (Because of the risk of poisoning as a result of misidentification, only pick the easily recognized edible species in the beginning.)*

Turkey tail fungi have zones of contrasting colour that actually look somewhat like a turkey's tail.
Drew Monkman

201

WEATHER

AUGUST HIGHLIGHTS

All Month Long

- Unlike June and July, we are now rapidly losing daylight. By month's end, the sun sets about 45 minutes earlier than it did on August 1.

- August is the second warmest month of the year in central and eastern Ontario.

- A familiar August smell is that of petrichor. This is the strong scent in the air when rain falls after a dry spell. The smell comes from a combination of soil bacteria and oils given off by certain plants during dry periods. The strong citrus smell of unripe green walnut fruits is another sign that summer is starting to wane.

- Summer heat and humidity are often accompanied by air quality warnings. One of the most important pollutants is ground level ozone, a colourless and odourless gas. When the Air Quality Index (AQI) for ozone climbs over 100, it causes serious respiratory effects. People with heart and lung disorders are at high risk.[5]

- Water levels often drop in late summer, revealing emergent shorelines. These sites are definitely worth exploring for many interesting plants such as the bladderworts.

Early August

- The hot weather of late July and early August is often referred to as the "dog days of summer." At this time of year, the "Dog Star," Sirius, rises in the southeast just before the sun. Because it is so bright, and its appearance often coincides with the hottest weather of the summer, this period became known as the dog days of summer.

Late August

- Heavy morning mists, especially in valleys and over lakes, complement the beauty of the August sunrise. With longer nights and lower temperatures, the night air is cooler in August. Water vapour condenses — and becomes visible — when it comes into contact with these cooler pockets of air. Coupled with the noticeably shorter days, they are yet another hint of impending autumn.

- Lawns and meadows are soaked with dew most mornings.

August Weather Averages, 1971–2000[6]

City or Town	Daily Max. (°C)	Daily Min. (°C)	Rainfall (mm)	Snowfall (cm)	Precipitation (mm)
Owen Sound	23.8	14.8	88.6	0	88.6
Huntsville	23.7	13.2	89	0	89
Barrie	24.8	14.2	92.6	0	92.6
Haliburton	23.4	12.2	85.5	0	85.5
Peterborough	25.1	13.4	91.6	0	91.6
Kingston	24	15	88.1	0	88.1
Ottawa	25	14.3	87.6	0	87.6

Approximate August Sunrise and Sunset Times (DST)[7]

(Note: Twilight starts about 30 minutes before sunrise and continues about 30 minutes after sunset.)

Location	Date	Sunrise (a.m.)	Sunset (p.m.)
Grey-Bruce	Aug. 1	6:10	8:51
	Aug. 15	6:26	8:31
	Aug. 31	6:44	8:04
Muskoka/Haliburton	Aug. 1	6:05	8:45
	Aug. 15	6:21	8:26
	Aug. 31	6:40	7:59
Kawartha Lakes	Aug. 1	6:00	8:40
	Aug. 15	6:16	8:20
	Aug. 13	6:34	7:54
Kingston/Ottawa	Aug. 1	5:53	8:32
	Aug. 15	6:08	8:13
	Aug. 31	6:27	7:46

Night Sky

August Highlights

All Month Long
- Major constellations and stars visible (August 15 — 10:00 p.m. DST)
  Northwest: Ursa Major with Ursa Minor (with *Polaris*) above to its right and Bootes (with *Arcturus*) to its left.

Northeast: Cassiopeia in mid-sky; Pegasus and Andromeda (with M31 galaxy) to its right; Perseus near the north horizon.

Southeast: Summer Triangle made up of *Vega* (in Lyra), *Deneb* (in Cygnus) and *Altair* (in Aquila).

Southwest: Sagittarius at the south horizon.

- The Algonquian name for the full moon of August is the Sturgeon Moon.

- The Perseid meteor shower peaks on August 12. (See below)

- The constellations Sagittarius and Scorpius boast numerous nebulae and star clusters that are visible with binoculars.

- By mid to late August, Orion is visible one hour before dawn on the eastern horizon.

METEOR MONTH

August 12 is usually the peak of the annual Perseid meteor shower, but viewing can also be good for a few nights before and after this date. Meteor watching is always best when there is little or no moonlight. However, even if there happens to be a bright moon, a single observer should still be able to see at least a dozen meteors per hour. The shooting stars originate in the Perseus constellation, low in the northeastern sky just below Cassiopeia. However, as the night progresses, Perseus climbs higher and higher in the sky and is nearly overhead by dawn. At that point, the meteors streak down in all directions toward the horizon.

A meteor shower occurs when the Earth passes through a stream of debris left behind by a comet. In the case of the Perseids, the meteors originate from debris left by the comet Swift-Tuttle. Most of the debris consists of particles the size of grains of sand which burn up and emit beautiful streaks light as they speed through our atmosphere at up to 160,000 kilometres per hour. Meteor showers are best viewed from a dark location as far away from city lights as possible. If you watch from a suburban backyard, you are likely to see less than 15 percent of the meteors visible from a dark site.

SEPTEMBER

Mists and Melancholy Joy

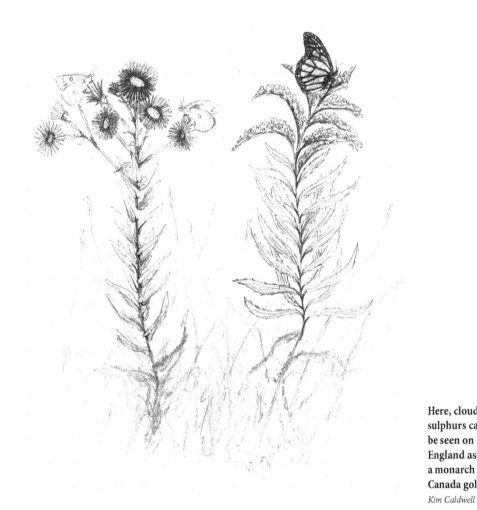

Here, clouded
sulphurs can
be seen on New
England aster, and
a monarch sits on
Canada goldenrod.
Kim Caldwell

Best I love September's yellow,
Morns of dewstrung gossamer,…
More than Spring's bright uncontrol
Suit the Autumn of my soul.

— Alex Smith

With September comes the beginning of fall, a season of melancholy joy and sentimentality. It is both a time of new beginnings and wistful endings. Almost without our knowing it, many of our migratory songbirds will slip away southward this month. Listening to their contact calls in the night sky, one feels a certain sadness at their departure but also a sense of wonder in the mystery of migration.

September is a month of golden yellows and plush purples as goldenrods and asters take over fields and roadsides. The late summer flowers are so abundant that September rivals May as a month of blossoms. Dawn is ushered in by heavy dews, mist, and sunshine softly filtered through countless spider webs. We awake to the raucous calls of Blue Jays and crows, the lisping of White-throated Sparrows, and the gentle notes of migrating warblers and vireos foraging in backyard trees and shrubs. During the day, the steady background chorus of crickets and katydids is punctuated now and again by the electric buzz of a cicada and the occasional call of a lone spring peeper. A walk through Shield country on a warm afternoon charms our noses with the scent of sweetfern and fallen pine needles, warmed by the late-summer sun. The citrus smell of green walnuts and the tang of windfall apples also tell us that September has arrived. Outdoor markets overflow with fresh fruit and vegetables that tantalize our taste buds as in no other month of the year.

On or about September 21, the sun crosses the equator on its annual southward course, marking the official beginning of fall. For the nature watcher, however, fall's approach has been felt since late July with the arrival of the first southbound shorebirds. In fact, the leaves of several species of trees will have almost reached their colour peak by the time the calendar gets around to acknowledging autumn's official arrival.

By the end of the month, the first flights of Northern Canada Geese will be going over and sparrows will have replaced warblers as the most common migrants. And then arrives the overwhelming colour. The stage has already been set by the colour show along roadsides, but the full symphony resides in the woodlands. Maples will set both city streets and country vistas ablaze with their oranges, reds, and yellows. And, no matter how many falls we witness, the vibrant colour of the leaves will never cease to astound us.

Birds

September Highlights

All Month Long

- September is to fall what May is to spring. Migration is in full swing and, given the right conditions, birding can be spectacular. Warblers, vireos, thrushes, swallows, flycatchers, and Rose-breasted Grosbeaks are some of the birds departing for Latin America in huge numbers. How they find their way to winter grounds is truly remarkable. (See page 208)

- Fall warblers challenge birders' identification skills. But they are actually much easier to identify than many people think. (See page 209)

- Even though some birds appear to migrate regardless of the weather, large movements of fall migrants usually occur just after the passage of a cold front. The northwesterly breezes provide tailwinds that facilitate flight and help the birds to conserve energy.

- American Robins continue to flock up. Strangely enough, they seem to take on the behaviour of a totally different species in the fall and often appear skittish, as if possessed by a restless urgency. Fall robins are seen less often on lawns but prefer to gorge themselves on berries.

- Large flocks of European Starlings, Red-winged Blackbirds, and Common Grackles continue to be widespread, especially in cornfields and wetlands. Increasingly rare Rusty Blackbirds are sometimes seen, as well.

- Check plowed fields for American Pipits, Killdeers, and even the occasional Black-bellied Plover, American Golden Plover, and Pectoral Sandpiper.

- A trip to the shores of Lake Ontario, Georgian Bay, Lake Huron, or the Ottawa River can be very rewarding from September through November. Songbirds, shorebirds, water birds, and hawks are all moving through in large numbers.

- Large numbers of Blue Jays make a mass exodus southward this month. Others will remain, however, to winter in central and eastern Ontario, especially in years when acorns and beech nuts are abundant. The Blue Jays' raucous calls are probably the most common bird sound of September.

- Although birdsong is almost entirely absent in the fall, some species will utter a half-hearted, tentative song on bright September and October mornings. Among those heard most regularly are American Robins, Purple Finches, and White-throated Sparrows.

Mid-September

- On sunny mid-September days when cumulus clouds dot the sky and the winds are from the northwest, watch for "kettles" (a group of birds wheeling and circling in the air) of migrating Broad-winged Hawks. Your best chance of seeing these birds is along the north shore of Lake Ontario.

- Large migratory flights of thrushes pass over about this time. Their loud, plaintive call notes are surprisingly easy to hear in the night sky, even over the city.

- Ruby-throated Hummingbirds abandon our feeders and surrender to the urge to migrate. Their southward flight includes a remarkable non-stop crossing of the Gulf of Mexico, taking 18 to 20 hours.

Late September

- White-throated Sparrows are moving southward in large numbers and can be a common sight at backyard feeders for several weeks.

- Large, high altitude flocks of Northern Canada Geese that nested along James Bay fly over as they make their way to wintering grounds in the Tennessee Valley. By adopting a *V*-formation, the lead bird breaks the initial air resistance, making flight easier for those birds following behind. However, with the lead bird using more energy than the others, the flock must change leaders frequently. Native peoples referred to late September and early October as "goose-going days."

- Large numbers of Turkey Vultures are often seen as they soar southward.

- Duck hunting season opens on or about September 25 for most of central and eastern Ontario.

- Ospreys leave cottage country for their wintering grounds in the southern United States, the West Indies, and Central and South America. Mangroves and coastal estuaries will be their home until next spring.

How Do Birds Find Their Way?

Sometime this month, just after darkness falls, a three-month-old Rose-breasted Grosbeak will depart its summer home in central and eastern Ontario and set its sights on Costa Rica. Although it has never made the trip before, it carries in its genes all of the necessary instructions to successfully navigate over such a huge distance. It will use senses and faculties that are no less than extraordinary. The grosbeak's almond-sized brain is able to make sense of crucial information coming from the stars, the sun, polarized light patterns, and even the Earth's magnetic fields. When one set of directional cues is obscured, as the sun and stars may be by cloud cover, more reliance is placed on alternate cues.

Even caged songbirds often become restless in the fall and will begin flitting against their cages just after sunset. Experiments with caged Indigo Buntings done in the 1960s showed that birds cue in to the stars that rotate around the North Star and that during their first spring and summer of life, they appear to memorize the position of these key stars in the northern sky.[1] To navigate by stars, birds require a clear view of the sky. However, birds often migrate below cloud level, which begs the question of how they find the proper direction. Researchers now have conclusive evidence that at least some migratory songbirds are able to orient themselves using the lines of the Earth's magnetic field that extend between the north and south poles. In 1984, it was discovered that the nasal tissues of Bobolinks contain magnetite. This magnetic mineral acts almost like a miniature compass needle. However, a detailed understanding of exactly how the birds use the magnetic field is still unclear. There is some evidence that birds may actually be able to see the magnetic field as a visual pattern or a specific colour.

When songbirds cannot rely on stars or the magnetic field for direction, they may turn to information from the position of the setting sun on the western horizon and/or the band of polarized light which extends perpendicular to the setting point of the sun. Invisible to humans, polarized light is created when sunlight scatters as it passes through the atmosphere. Just as the sun location changes with latitude and the time of year, so does the position of the band of polarized light. These cues can therefore be used by birds in choosing their bearings. Even wind-carried odours may provide important directional information. So, it may be that there is something in the Rose-breasted Grosbeak's genes that attracts it to the smell of Central American lowland rainforest! The mystery of migration continues.

The Rose-breasted Grosbeak's rich, warbled song is often compared to a "robin who has taken singing lessons."
Karl Egressy

IDENTIFYING THE FALL WARBLERS

Many birders enjoy the challenge of identifying the fall warblers. In some ways, warbler-watching is actually better in the fall than in the spring. The daily stream of birds is steadier and less dependent on the idiosyncrasies of the weather. The migration period is also much longer, extending from late July, when the first Northern Waterthrushes and Yellow Warblers migrate through, until early November, when the last of the Yellow-rumped Warblers are seen.

When Roger Tory Peterson published his *Field Guide to the Birds*, he unfortunately set a negative tone regarding the identification of fall warblers. On two pages entitled "Confusing Fall Warblers" he presented the dingiest, drabbest, and least remarkable of these birds, leading some people to believe that the entire warbler assemblage is exceedingly hard to identify after the end of June. Such, however, is not the case. Some species do not change at all (e.g., American Redstart) and many are just slightly altered from their spring plumage (e.g., Black-throated Green Warbler). Immature birds present the biggest challenge, although one, the Chestnut-sided Warbler, is so distinctive that it is easier to identify than some spring adult species. As a general rule, adult warblers migrate earlier than the immatures. By mid-September, the majority of warblers seen tend to be young birds. The secret to enjoying fall warbler-watching is not to worry about identifying everything you see. Just let some of the LYJs (little yellow jobs!) flit on by. You will see enough that is familiar to come up with a very respectable species total. Be sure to use the pishing technique, however. Warblers are very inquisitive.

MAMMALS

SEPTEMBER HIGHLIGHTS

All Month Long

- The members of the squirrel family are especially active and conspicuous in late summer and early fall. (See page 211)

- Little brown and big brown bats mate and take up residence in hibernation sites. Other bat species such as red bats migrate south. (See page 211)

- White-tailed deer are feeding heavily to build up fat that will supply up to a third of their winter energy needs. A grey-brown winter coat replaces the reddish summer coat in September. With dense inner fur and long, hollow outer hairs, it is actually ten times thicker than the summer coat. The grey-brown coloration will offer excellent camouflage in the winter woods.

- Bucks expend considerable time this month rubbing their antlers against branches in order to peel off the velvet covering. The main function of the rubbing behaviour, however, is to deposit scent. The scent, which originates from glands in the forehead, serves to advertise both the presence and rank of the buck. Only the largest and most dominant males will have the opportunity to mate with the does.

- Sensing the shorter days, beavers begin cutting down trees once again for winter food. Aspens are the preferred species.

- During the autumn months, a bear may feed up to 20 hours a day, consuming 20,000 calories of food in the process. This can result in a weight gain of over 45 kilograms.

Late September

- Moose mate from late September to late October. Being such romantic beasts, they time the rut to coincide with the height of the fall colours. Rutting (behaviours and activities related to mating, including encounters between males) usually takes place in the early morning or late afternoon.

A BUSY TIME FOR SQUIRRELS

September is a time of heavy feeding for many species of mammals as they prepare for the coming winter. The eastern chipmunk is one of the most noticeable mammals at this time of year. Its staccato "chuck-chuck-chuck" is a common woodland sound as it collects nuts and seeds to store in underground pantries. The chipmunk is not a true hibernator and therefore must visit its pantries over the course of the winter. Dietary staples include hazelnuts, beech nuts, and acorns, as well as seeds from conifers and herbaceous plants. Berries and mushrooms are also eaten in large numbers.

The shortening daylight hours also trigger a change in the behaviour of gray squirrels. Their activity, too, becomes dominated by eating and storing food such as beech nuts, walnuts, and acorns. To store food, gray squirrels dig a shallow hole or "cache," deposit a single food item, and then cover it with soil. Relocating buried food is based on the squirrel's excellent spatial memory, nearby landmarks, and the animal's keen sense of smell when it is within a few centimetres of the cache. Frequently, however, the food a squirrel finds was actually buried by another squirrel.

As for red squirrels, they're busy gathering cones, especially from spruces and hemlocks. The seeds in the cones will ripen in storage and be ready for winter consumption. They will also place apples or mushrooms on the fork of a twig to ripen or dry and be eaten at a later date.

BATS ON THE MAKE AND ON THE MOVE

On early fall evenings, little brown and big brown bats are sometimes seen feeding in small groups or moving toward their wintering destinations. Both species usually spend the winter months in Canada, although they do sometimes fly considerable distances to hibernation sites. Three other bat species that hibernate in Ontario, albeit in smaller numbers, are the northern long-eared, the eastern pipistrelle, and the eastern small-footed. Some Ontario species such as red, hoary, and silver-haired bats migrate south to the Gulf States. Migratory movements of these bats can sometimes be seen at Great Lake locations such as Prince Edward Point.

Male and female little brown and big brown bats swarm at hibernation sites in late summer before actually beginning hibernation at the end of September or early October. Mating activity is believed to be the main purpose of the swarming phenomenon. It is interesting to note that, after mating, the sperm is stored in the female's body until spring, when ovulation and fertilization take place. From September to May, people should stay out of any caves or mines where bats are hibernating, as disturbance costs the animals' energy and is usually fatal.

Amphibians and Reptiles

September Highlights

All Month Long

- Leopard frogs, in particular, are starting to move from summer feeding sites such as meadows and marshes to their hibernation sites in streams, ponds, and rivers. Some will travel considerable distances. Amphibians prefer to travel during warm, wet weather, especially at night. If conditions for migration have been too dry for a long period of time but then suddenly become suitable, the numbers of frogs on the move can be quite spectacular. Where these migrations involve crossing busy roads, large numbers of frogs are killed.

- Although their reproductive purpose is for another season, spring peepers sometimes make half-hearted, sporadic calls from woodland trees in the fall. They are most vocal on warm, damp days. It is rare to hear more than one or two calling at a time, however.

- Red-backed salamanders mate in the fall. The male deposits a spermatophore — a white, pyramid-shaped mass of sperm-filled mucus — that the female picks up with her cloaca to fertilize the eggs. The eggs will be laid in the spring.

Fish

September Highlights

All Month Long

- Muskellunge and Walleye move to shallower water and feed heavily. Females of both these species need to eat a great deal in the fall in order to provide nourishment for the eggs developing in their bodies. Late September through November offers the best fishing of the year for these species. Walleye, however, is in decline in central and eastern Ontario. (See below)

Late September

• Brook Trout also feed heavily and start moving upstream in preparation for spawning. They can sometimes be seen stopping to sun themselves over submerged rocks or logs. Males begin to acquire their deep orange to red nuptial colours.

WALLEYE DECLINE

There is concern about the future of Walleye populations in many parts of central and eastern Ontario. Both the size and number of Walleye caught has declined. A number of factors are working against these fish. One major stressor is the extremely high level of angler harvest. More than 25 percent of all fishing in southern Ontario is directed at Walleye and it remains the most highly sought fish across the province. Central and eastern Ontario also has fewer large adult females than in areas farther north. This is often associated with stressed or unhealthy Walleye populations.

Another major concern is the impact of introduced or invasive species, which can greatly reduce the amount of Walleye a given lake can support. Because Walleye are adapted to low to moderate light conditions, these fish do not do well in lakes with high water clarity. Zebra mussels, filter-feeders that are now present in many parts of central and eastern Ontario, including all of the Kawartha Lakes, are making our lakes increasingly clearer and thereby decreasing their suitability as Walleye habitat. Another invader, the Black Crappie, competes directly with adult Walleye for food and often preys directly on young Walleye.

Shoreline alteration is another threat. The construction of docks, boat houses, and retaining walls has the cumulative effect of drastically altering many shorelines and, in doing so, adversely affecting the spawning, nursery, and feeding habitats of near-shore fishes such as Walleye. Finally, climate change is also working against Walleye populations. A warmer climate will reduce the amount of cool water habitat available for the fish, which will make many central and eastern Ontario lakes less suitable for this popular species. To help address some of the concerns regarding Walleye, there are now new limit and size regulations.[2]

Insects and Other Invertebrates

September Highlights

All Month Long

- Large mating swarms of ants are a common September phenomenon, especially on warm, humid afternoons. Watch for them even in the city. (See page 216)

- Spider webs are everywhere. They are especially visible in the early morning on shrubs and grasses near wetlands. (See page 217)

- Goldenrods and asters attract huge numbers of insects and provide excellent opportunities for close-up observation and photography. Look for honey bees, wasps, long-horned beetles, soldier beetles, ambush bugs, monarch butterflies, and mantises, as well as hover flies, which mimic bees or wasps in appearance.

- Migrating monarch butterflies are on the move as they make their way south to the Great Lakes and then on to the mountains of western Mexico where they will spend the winter. Many will go to the El Rosario Sanctuary near the town of Angangueo.

- The almost total absence of biting insects makes outdoor activities especially pleasant.

- The webs of the fall webworm (*Hyphantria cunea*), a member of the tiger moth family, stand out noticeably. The large, loose webs encase the ends of the branches of broad-leaved trees and house colonies of beige caterpillars that will become white moths in the spring. They are mainly an aesthetic pest and rarely cause the tree any serious harm. The webs often remain on the tree throughout the fall and winter and are easy to mistake for tattered old oriole nests.

- Cuddly brown and black woolly bear caterpillars are a common sight as they look for a sheltered location to overwinter. According to folklore, the wider the caterpillar's middle brown band, the milder the winter will be; studies trying to link wooly bear band width and the severity of the coming winter show no relationship exists, however. Watch also for yellow bear and American dagger moth caterpillars, which are similar in size and also have a hairy appearance. The latter is a white caterpillar with five black hair "daggers" emerging from its back.

- Dragonflies are still fairly common, especially small yellow-legged and white-faced meadowhawks and some of the darners. Several species of spreadwing damselflies can also be seen until well into October.

- Part of the common green darner dragonfly population migrates to the southern United States in the fall. They are capable of flying well over 100 kilometres a day!

- Hornets and yellowjackets continue to be abundant. The workers (sterile females) spend less time at the hive feeding the young in early fall and more time looking after their personal needs for sugar and sweet liquids.

- A number of freshwater mussel species spawn in early fall. Mussels have beautifully shaped and detailed shells as well as a fascinating reproductive cycle. This includes using fish as a host when the mussels are in the larval stage. Unfortunately, mussels are among the most endangered organisms in North America.

Early September

- Clouded sulphur butterflies reach their peak numbers at this time of year and are usually the most commonly seen butterfly species. Other widespread late summer and early fall species include cabbage white, meadow fritillary, pearl crescent, viceroy, common ringlet, eastern comma, and monarch.

- The spiralling flight of pairs of white or sulphur butterflies is a commonly seen behaviour during the mating season. A male and female butterfly will circle around each other, all the while ascending high into the sky. Then, without warning, the male will give up the chase and drop to the ground, almost like a dead weight. The female then slowly glides back down. It is believed that the female initiates these aerial climbs to rid herself of unwanted suitors, probably because she has already mated.

Unlike early and mid-summer monarchs, which only live for a few weeks, the late-summer generation migrates south to Mexico and lives for seven or eight months.

Terry Carpenter

- Monarch butterfly numbers are at their highest. The monarchs begin to congregate at peninsulas on the Great Lakes such as Presqu'ile Provincial Park, a jumping off point for their migration across Lake Ontario and on to Mexico. A monarch tagging demonstration is held here each year on Labour Day weekend and is well worth attending. A tagger is on hand to answer questions about monarchs and to show how the butterflies are tagged with a tiny adhesive sticker bearing a number and return address.

- Although the song of the cicada fades away early in the month, crickets, katydids, and grasshoppers continue to call fervently during the day. As temperatures become cooler, however, the amount of evening song decreases noticeably.

SWARMING ANTS

On hot, muggy days in late summer, we are sometimes provided with an intriguing glimpse into the lives of ants. This is when we see spectacular swarms of "flying ants" milling about on sidewalks or flying high overhead. Ants only have wings during the mating phase of their life cycle. Mating takes place over the course of just one day in a phenomenon known as swarming. Swarming behaviour most often occurs in the afternoon, usually a few days after a heavy rain, when the weather is warm and humid. Thousands of winged males and much larger winged females emerge from the ground, usually pushed out by wingless workers. The male ants are attracted to powerful sex pheromones produced by the female. Copulation often occurs on the wing and a queen will mate with several males. Swarming behaviour is usually synchronized with other ant colonies of the same species so that it occurs on the exact same day over a large area and promotes interbreeding.

An especially interesting behaviour of mating swarms in some species is known as "hilltopping." In what is probably an effort to more easily find a mate, ants congregate around prominent points of a landscape such as a large tree or a chimney. When ants are swarming, it is not uncommon to see predators such as dragonflies or even Ring-billed Gulls grabbing the ants out of mid-air and quickly gobbling them up. When the swarming ceases, all of the male ants die, and each mated female begins the very difficult task of establishing her own new colony. The first thing she does is to shed her new wings by biting them off. You can often see wings lying about after swarming has occurred. The wing muscles, which are no longer needed for flight, become an important source of nutrients during the initial stages of colony development.

Most of the ants we commonly see in central and eastern Ontario are formicine ants belonging to the genus *Formica*. They are typically black or black and red and are often seen on sidewalks. Formicine ants have a distinctive node (prominent bump) on the constricted waist between the thorax and the abdomen. However, you can only see it well with a good hand lens. Another way to identify them is to simply pick up one or two. You should be able to get a good whiff of formic acid. This chemical serves a number of purposes ranging from communication to defence. One of the most common formicine ants is the black carpenter ant. They are infamous for damaging wood structures when excavating nests.

Ants are hugely important ecologically. Not only are they the premier soil turners but they are also the key to the success of many of our spring wildflowers. A small pouch containing a sweet liquid is attached to the seeds of early blooming flowers such as trilliums and violets. Ants find this liquid irresistible and hoard the seeds in their underground chambers. Once there, the seeds quickly germinate in the moist, dark soil.

Ubiquitous Spiders

Late summer is a special time of year. Dawn arrives with heavy dews, gentle mists, and a sort of melancholy peace. It is on these quiet beginnings to the day that the interaction of early morning sunlight and lingering dew can reveal dozens if not hundreds of spider webs. Spiders are carnivores belonging to the class Arachnida, which also includes animals such as scorpions, mites, ticks, and harvestmen (daddy longlegs). All spiders produce venom with which they subdue their prey. However, the only species native to Ontario that can be dangerous is the northern black widow. Fortunately, it is very rare here, and extremely secretive and docile.

Central and eastern Ontario is home to hundreds of species of spiders of many different families. Surprisingly, only a few of these families build webs and it is always the female who does the building. Probably the most noticeable webs at this time of year are those of a family known as the orb weavers. These are the classic vertical webs attached to surfaces such as branches, buildings, and fences. On early morning walks, you can often see hundreds of these webs strung between plants, especially in brushy swamps. The rays of silk that pass through the centre have no adhesive quality, thereby allowing the spider to walk on them. Insects are caught in the sticky, more fragile spiral threads. These threads are laden with microscopic droplets of an insect adhesive or "tangle-foot." The webs are often constructed at night. One orb weaver that is particularly conspicuous at this time of year is the black and yellow argiope (*Argiope aurantia*). This large spider often builds its web in a clump of goldenrods and hangs conspicuously right in the hub.

Walking across a lawn or grassy meadow, keep an eye out for flat, horizontal webs with a funnel-like tube at one edge. They are built by funnel weaver spiders. The grass spider (*Agelenopsis potteri*) is a common member of this group. Lurking in the funnel, the spider dashes out, grabs its prey, and quickly returns to the funnel to eat. Another common late summer Arachnid is the goldenrod spider (*Misumena vatia*). It is a small white or yellow species belonging to the crab spider family. The manner in which it holds its legs out at the sides and quickly moves sideways is reminiscent of a crab. Especially fond of goldenrods, it sits hidden in the flowers with almost perfect camouflage. When a bee or hover fly comes too near, the spider clamps it with its powerful legs and delivers a paralyzing bite. The best way to find these spiders is to look for a dead bee on a flower.

The black and yellow argiope, a common orb spider, spins a circular web that can measure 60 centimetres across.

Terry Carpenter

PLANTS AND FUNGI

SEPTEMBER HIGHLIGHTS

All Month Long

- The leaves of most native trees and shrubs will begin to change colour this month. Understanding how and why this happens adds a great deal to our enjoyment of fall. (See page 220)

- Some species of trees can be easily identified in the fall based solely on the colour they have turned. For other species, colour-based identification is more difficult, since they may turn one of several colours. (See page 222)

- The time and order in which the various tree species reach their colour peak is surprisingly consistent from one year to the next. (See page 222)

- The amount of fruit — seeds, berries, acorns, keys, and the like — on trees and shrubs varies considerably from one year to the next. In a masting year, the fruit crop can be especially abundant. (See page 222)

- Goldenrods and asters turn fields a riot of yellow, purple, and white. Goldenrods dominate early in the month, but asters reign supreme by month's end. (See page 223)

- Gentians, a radiant blue flower of late summer, are in bloom. Other September wildflower specialties include false dragon-head, grass-of-Parnassus, Kalm's lobelia, gerardia, white snakeroot, and ladies' tresses orchids.

- Vines and shrubs put on a wonderful colour display this month. Sprawling over fences and spiralling up dead trees, Virginia creeper glows with scintillating reds, as do strawberry and blueberry leaves. Poison ivy offers up lovely oranges, while dogwoods and blackberry bushes provide beautiful burgundies.

- The vines of riverbank grape, Virginia creeper, wild cucumber, and virgin's bower stand out prominently along roads and trails. Their numerous branches and tendrils enmesh just about every available shrub, fence, and fallen tree. Virgin's bower can be identified by the grey, silky plumes of the flowerheads. Wild cucumber has roundish, cucumber-like seed pods covered in soft bristles.

- A large variety of native grasses, sedges, and rushes that bloomed in the summer can actually be best identified in the fall. This is because seeds rather than flowers are often the most important features for identification. The fall is a wonderful time to study this generally neglected part of our flora.

- White pine cones become mature and most of the seeds are released. Red pine, white cedar, white spruce, tamarack, and balsam fir also begin to free their winged seeds to the wind. Fir cones are unique in that they stand straight up like candles. Their scales don't just open up to release the seeds but actually fall right off. Soon, all that remains of each fir cone is its stick-like core. Look for them in the top of the tree.

- Mushroom diversity and abundance is usually very good. You may even smell the fetid odour of a stinkhorn mushroom, which can "scent" an entire backyard. Also watch for giant puffballs, an edible species that often grows in fields.

Mid-September

- Red maples growing along the edges of lakes and wetlands reach their peak colour by mid-September or earlier.

- Even before the first frosts, cattail leaves begin to turn brown, as do bracken ferns. But before the fronds begin to shrivel, bracken glows in the September sunshine in a unique blend of brown, yellow, and gold.

- Beechdrops bloom under beech trees this month and next. A parasitic species, beechdrops do not carry out photosynthesis but receive their nourishment from the roots of the beech. The small, tubular flowers are white and purple.

Late September

- Oaks are now shedding their acorns. They are gobbled up by all manner of birds and mammals.

- By late September or the first week of October, the maples of the Canadian Shield and Algonquin Park are usually close to their colour peak. Consider taking a "colour drive." (See page 224)

- In years of little rainfall, colour change and leaf drop can occur much earlier than usual. Trees will shed their leaves early in order to conserve water, which is continually lost through the leaves by transpiration.

- The purples, mauves, and whites of asters now reign supreme in fields and along roadsides.

- The appearance on lawns of the shaggy mane mushroom (*Coprinus comatus*) is a sure sign of fall. The brown or white cylindrical caps are very distinctive.

- Old needles die and are shed from white cedar and white and red pine. Fallen pine needles often accumulate in large piles on roads and sidewalks.

- Most years, white ash, pin cherry, and staghorn sumac reach their colour peak about now. Some ash trees turn a stunning purple-bronze that glows in the September sun.

WHY THE COLOURS?

There is a beautiful Native American legend that talks of hunters in the north sky who killed the Great Bear — represented by the constellation bearing the same name — in autumn, and its blood dripped down over forests, colouring the maples red. Later, as they cooked the meat, fat dripped from the heavens, turning the leaves of the aspens and birch yellow. Quite clearly, the fall colours have never ceased to amaze human beings and to make us wonder why they appear. In a nutshell, colour change and the shedding of leaves are manifestations of a tree's preparation for winter. Since winter is a time of drought, when water is locked up in the form of ice, trees are no longer able to take up water through their roots. Because leaves are continually releasing water vapour — think of the high humidity of a greenhouse — trees must get rid of their leaves in order to minimize water loss and desiccation. However, if the leaves just froze in place, snow and ice would build up on the foliage and break off entire limbs. The tree would also lose the minute but precious quantities of minerals concentrated in the leaves that were originally obtained through its roots. Magnesium in particular is essential for the production of chlorophyll, the green pigment in the leaves. Through the process of photosynthesis, chlorophyll captures the sun's energy and uses it in combination with water and carbon dioxide to produce the sugar-based substances that make up a tree's tissues. A complex process has therefore evolved to salvage as much of the mineral content from the leaves as possible and store it in the tree's woody tissues before the leaves are lost. The nutrients will then be readily available to help manufacture a whole new set of leaves come the spring.

As summer progresses and the leaves produce less and less chlorophyll, colour change slowly becomes apparent. With less and less chlorophyll there to conceal them, other pigments in the leaves gradually become visible. The yellows and oranges come from carotene pigments which have been present in the leaves all along. The stunning reds and purples, however, are produced by anthocyanin pigments that are created by excess sugars in the leaf. These pigments seem to be brightest in years when there is lots of late summer sunshine and consistently cool — but not freezing — nights. Frost kills the leaf's tissues and puts an end to the chemical processes that result in good colour production. An extended summer drought runs counter to good colour, as well. Because the trees lack water for photosynthesis, they can't produce the sugars needed for intense red colours. An extremely wet fall will also cause muted colours.

The actual shedding of the leaves is achieved by the formation of a layer of "abscission" cells between the stem and the twig. Coming from the same root as the word *scissors*, these cells are designed to make a "cut" by pushing the leaf, little by little, away from the twig. Eventually, they also make it impossible to transfer in the minerals the leaf needs to make more chlorophyll. In the end, the leaf's connection with the twig is broken and it falls off in the wind, rain, or simply from the warming effect of the morning sun. At this melancholy time of year, it is comforting to think of the fall colours and leaf drop not as the demise of this year's leaves, but rather as the transfer of precious nutrients back to the parent tree to use in next spring's foliage.

The colours of sugar maple leaves can range from bright yellow to red-orange.
Terry Carpenter

221

WHAT SPECIES TURNS WHAT COLOUR?

For many trees, such as tamaracks and trembling aspen, fall colour is predictable; they both turn yellow. For other species, there is much more variability. White ash, for example, can turn any colour from bronze-yellow to wine-purple. Male red maples usually turn red, while female trees tend to be yellow. The following list outlines the most common colours that leaves become before being shed.

Reds, purples, and burgundy: Red maple, white ash, pin cherry, staghorn sumac, blackberry, red raspberry, blueberry, red oak, bur oak, dogwoods, Virginia creeper.

Pinks: Maple-leaved viburnum, red maple, chokecherry, dogwoods.

Oranges: Sugar maple, red maple, red oak, chokecherry, staghorn sumac, bigtooth aspen, poison ivy.

Yellows: Tamarack, sugar maple, silver maple, red maple, Norway maple, trembling aspen, bigtooth aspen, white birch, balsam poplar, striped maple, American beech, bitternut hickory, black walnut, white ash, American basswood, tamarack, American elm, willows, purple-flowering raspberry, riverbank grape.

Browns and coppers: American beech, white ash, red oak, bur oak, bracken fern, white cedar, white pine.

Remain green: Speckled alder, common lilac, European buckthorn.

THE COLOUR TIMETABLE

The following are the approximate dates when some of the better-known species usually reach their colour peak. There is always variation, however, depending on the weather and the trees' exposure to sunlight.

Late September: White ash, pin cherry, staghorn sumac.

Early October: Red maple, sugar maple, black walnut, American basswood, American beech, American elm.

Mid-October: Red oak, bur oak, bitternut hickory, silver maple, trembling aspen, bigtooth aspen, balsam poplar.

Late October: Tamarack, Norway maple.

BOOM OR BUST IN THE TREES

Not surprisingly, millions of years of evolution have fine-tuned plants as much as animals to be able to survive and pass on their genes to a new generation in a rough and tumble world. For a tree, the challenge is how to best avoid having all of its seeds eaten by predators. Therefore, any adaptation that minimizes such predation of the seeds will be favoured by natural selection. One of the things we notice in trees is the marked difference in the amount of seed — berries, nuts, acorns, cones, keys, etc. — produced from one year to the next. In fact, many tree species will produce relatively little seed for several years in a row. This has the effect of greatly reducing the populations of seed predators because, with little or no appropriate seed around, many will die. Others may leave the area to search for food elsewhere or will simply have fewer young. Then, with the predator population knocked down a few notches, something amazing happens: the trees suddenly produce a giant crop of seeds. Those predators that are still around quickly become satiated and can eat no more. In this way, a significant number of seeds will survive to grow into fledgling trees. The following year, most trees return to their stingy ways and produce few if any seeds. Just when animals thought they had it made and produced much larger than usual families, the food dries up! Talk about trickery and deceit. Clearly, a parent tree doesn't need to produce new seed every year, because the young trees it has already produced still lay waiting for their turn to get enough sun and other resources to put on a growth spurt and maybe even grow to maturity.

In this boom or bust cycle of seed production, the manufacture of prodigious quantities of seed is known as masting. Different species of trees will often have a mast year at the same time and over a large geographic area. In 2008, it was the white pines that, across central and eastern Ontario, conspired to produce a huge cone crop. In 2011, it was the sugar maples and white cedar. How trees manage to coordinate the same cycle of seed production over such huge distances is still unclear, although some researchers believe that trees may have biological clocks that are somehow synchronized and pre-programmed to mast at opportune times. Masting has important ecological effects, too. Deer mice population cycles, for example, are closely tied to seed production in sugar maples. As the seed production of maples goes up and down so wildly from one year to the next, the mouse population follows suit. Since a number of larger birds and mammals depend on mice as a major source of food, the fluctuations in sugar maple seed production have a major ecological ripple effect through the wildlife community.

ASTERS AND GOLDENRODS

Fall is the season of asters and goldenrods. These underappreciated native plants are, in many ways, showier than the flowers of spring. Central and eastern Ontario is home to about 17 species in each of these plant groups. Although many of the asters are similar in appearance, knowing some of the more distinctive species adds interest to any fall outing. However, because the common names often change from one field guide to another, it is essential to also know the scientific names (although they sometimes change, too!) In fields and along roadsides, watch for the white flowers of the heath (*Symphyotricum ericoides*), panicled (*S. lanceolatum*), and calico asters (*S. lateriflorum*) as well as the violet-coloured blossoms of smooth (*S. laeve*), heart-leaved (*S. cordifolium*), and New England asters (*S. novae-angliae*). In damp thickets, purple-stemmed aster (*S. puniceum*) is quite common along with flat-topped white aster (*Doellingeria umbellata*). In woods and other shaded areas, large-leaved aster (*Eurybia macrophylla*) is common. The true star of the aster parade, however, is the New England aster. The colour contrast between the rich, violet rays and the deep orange-yellow centre make this plant stand out like few others.

As for the goldenrods, the most common species in dry, open areas are the early (*Solidago juncea*), grass-leaved (*Euthamia graminifolia*), stout (*S. squarrosa*), and especially Canada (*S. canadensis*) goldenrods. In rich, open woods, look for blue-stemmed (*S. caesia*) and zigzag (*S. flexicaulis*) goldenrods. Bog goldenrod (*S. uliginosa*) is a common species of wetland borders. Many of the asters and goldenrods have different blooming periods and some asters are still in flower in mid-October. The particular mix of species also changes between southern central Ontario and Shield areas. The Ontario Wildflowers website (see List of Websites) is especially good for identifying asters and goldenrods.

SEEING THE FALL COLOURS

For most people, seeing the fall colours means seeing the red and sugar maples. They usually reach their peak by the first week of October in the more northern parts of central and eastern Ontario and a week or so later farther south. Remember, however, that the maples are not the whole show. The oaks, with their stunning burgundies and browns, and the aspens with their flaming yellows, also provide a beautiful display a little later in the month. Throughout the fall colour season, a province-wide information line provides twice-weekly updates on the progression of the colours. For up-to-date information call 1-800-Ontario or visit *www.ontariotravel.net/fallcolourreport*.

Although there is spectacular colour to be seen just about everywhere, some of the more spectacular colour drives include:

Kawarthas: From Apsley, travel east along County Road 504 to Lasswade and on to County Road 620. Turn west to Glen Alda and back to Highway 28.

Muskoka: Bracebridge east on County Road 117 to Baysville and then on to Dorset, where you can climb the fire tower. Continue north through Dwight and then follow County Road 9 south back to Baysville or take Highway 60 through Algonquin Park.

Haliburton and Madawaska: From Haliburton Village, go east to Wilberforce, north to Harcourt, and continue north on County Road 10 to Maynooth. From Maynooth, continue east on County Road 62 to Barry's Bay. There is a scenic lookout just before Combermere.

Georgian Bay: On the Beaver Valley route, take County Road 7 south from Meaford, with a stop at Epping for spectacular views of the Niagara Escarpment. Continue south to Kimberley, Eugenia Falls, and Flesherton. You can return via Kimberley and drive north on County Road 13, following the Beaver River to Thornbury.

Land O' Lakes: Highway 41 from Kaladar north to Griffith (and beyond) is very scenic. You may wish to also take County Road 71 (Matawatchan Road) east along the Madawaska River.

Kingston: From Kingston's Division Street, travel north on County Road 10 (Perth Road) to the village of Westport. From Westport, travel northwest toward Maberly on County Road 36, then drive west along Highway 7 to Sharbot Lake. Return south on Highway 38.

Ottawa Valley: Starting at Castleford, north of Arnprior, take County Road 20 west to Renfrew and continue on Highway 132 to Dacre. The "Opeongo Line" begins about one kilometre north of Dacre. Drive west on County Roads 64, 512, and 66 to Highway 60 and Barry's Bay.

Lake Ontario: From Grafton, travel along Highway 2 through Brighton to Trenton and proceed south on Highway 33 to Picton, where you can visit Prince Edward County.

WEATHER

SEPTEMBER HIGHLIGHTS

All Month Long

- Daylight continues to decrease rapidly. The month begins with more than 13 hours of daylight but ends with less than 12.

- Temperatures become noticeably cooler. The daily average temperature for the month is about five degrees lower than in August.

- Sun-warmed sweetfern leaves and fallen pine needles scent the air of Shield country, especially on granite outcroppings. The smell of the first fire in the woodstove on a cool cottage country morning also announces the change of season.

- Precipitation across central and eastern Ontario is well above the monthly average.

- Heavy morning mists dance and curl over rivers, lakes, and valleys. Land and water surfaces that have warmed up during the summer are still evaporating a lot of water into the atmosphere. The moisture condenses into water droplets when it comes into contact with the cool early-morning air. The combination of mist, scintillating leaves, and the rising sun give September dawns a beauty unequalled at any other time of the year.

- September only rarely brings the high humidity and oppressive heat of July and August. For many of us, it is the month we shake off our summer lethargy and come back to life.

Early September

- Despite what the calendar is saying, it already looks and feels like fall. Daylight and darkness are now almost equal in duration. Already, dusk is two hours earlier than it was at the end of June.

Late September

- The fall equinox takes place on or about September 21. We are now losing daylight at the rate of about three minutes a day. For the next six months, nights will be longer than days.

- On or about September 26, day and night are almost exactly equal in duration.

- The first subfreezing temperatures since the spring are usually recorded along with the first frost. Low-lying areas are especially vulnerable to frost.

SEPTEMBER WEATHER AVERAGES, 1971–2000[3]

City or Town	Daily Max. (°C)	Daily Min. (°C)	Rainfall (mm)	Snowfall (cm)	Precipitation (mm)
Owen Sound	19.6	11	105.2	0	105.2
Huntsville	18.7	8.8	105.1	0	105.1
Barrie	20.1	9.6	97.6	0	97.6
Haliburton	18.4	8	86.6	0	86.6
Peterborough	20.1	9	84.3	0	84.3
Kingston	19.5	10.4	93	0	93
Ottawa	19.7	9.7	86.8	0	86.8

APPROXIMATE SEPTEMBER SUNRISE AND SUNSET TIMES (DST)[4]

(Note: Twilight starts about 30 minutes before sunrise and continues about 30 minutes after sunset.)

Location	Date	Sunrise (a.m.)	Sunset (p.m.)
Grey-Bruce	Sept. 1	6:45	8:03
	Sept. 15	7:02	7:37
	Sept. 21 (equinox)	7:09	7:25
	Sept. 30	7:19	7:09
Muskoka/Haliburton	Sept. 1	6:41	7:57
	Sept. 15	6:57	7:32

	Sept. 21 (equinox)	7:04	7:20
	Sept. 30	7:15	7:04
Kawartha Lakes	Sept. 1	6:35	7:52
	Sept. 15	6:51	7:26
	Sept. 21 (equinox)	6:58	7:15
	Sept. 30	7:09	6:58
Kingston/Ottawa	Sept. 1	6:28	7:44
	Sept. 15	6:44	7:19
	Sept. 21 (equinox)	6:51	7:08
	Sept. 30	7:02	6:51

NIGHT SKY

SEPTEMBER HIGHLIGHTS

All Month Long

- Major constellations and stars visible (September 15 — 9:00 p.m. DST)
  Northwest: Ursa Major with Ursa Minor (with *Polaris*) above, and Bootes (with *Arcturus*) below to left.
  Northeast: Cassiopeia in mid-sky with Perseus to its lower right; Pleiades just above horizon.
  Southeast: Pegasus (with the Great Square) and Andromeda (with M31 galaxy).
  Southwest: Summer Triangle made up of *Vega* (in Lyra), *Deneb* (in Cygnus), and *Altair* (in Aquila); Sagittarius near south horizon.

- The Harvest Moon — the full moon closest to the fall equinox — usually occurs in September. For several days the moon rises close to the same time every evening. (See page 228)

- The Algonquian name for the full moon of September is the Harvest Moon.

- At the fall equinox, both the moon and sun rise due east and set due west.

- This is still an ideal time to explore the Milky Way. Autumn nights are some of the clearest of the year, it is dark early, and in the early evening the Milky Way is right overhead.

- Pegasus and its asterism, the Great Square, is the best-known constellation of the fall. We can also see our galactic neighbour, Andromeda, in this part of the sky. It appears through binoculars like a faint oval of fuzzy light — light that left the galaxy 2.3 million years ago! Andromeda reminds us that, when we look into the night sky, we are looking into a gigantic time machine and seeing astronomical events that took place eons ago.

THE MOONLIT EVENINGS OF HARVEST TIME

Long before we had modern calendars, people gave fanciful names to the full moons of the year to help keep track of time. The Harvest Moon was — and still is — the name given to the full moon closest to the fall equinox or first day of autumn. It can therefore occur before or after the equinox. Anyone familiar with the comings and goings of the moon knows that the full moon rises in the east just as the sun is setting. This is because the full moon is always opposite the sun in the sky. On average, the moon rises about 50 minutes later on each subsequent day. Moonrise times around the fall equinox, however, do not follow this general rule. In September and October, the full moon rises an average of only 25 minutes later for several evenings in a row and seems to linger above the horizon as it follows a shallow angle up into the sky. The moon appears full or nearly full on all of these nights. This means, of course, that our evenings are lit up by moonlight. To busy farmers, these moonlit evenings are still a much-appreciated bonus of light. They allow the equivalent of two additional days of harvesting, hence the name Harvest Moon.

The Harvest Moon, like other full moons during the year, also appears to be larger at moonrise than when it is riding high in the sky. This isn't because the moon is any closer than usual. It is simply an illusion. When measured or photographed, the moon is exactly the same size, no matter where it is in the sky. Our eyes have much more experience judging the size of objects located straight ahead. We also tend to relate the size of the rising moon to the hills, trees, and buildings that appear to be close by it. On the other hand, we tend to see things situated high above us as being smaller. This is also

The Harvest Moon — or any other full moon — at the horizon only looks big because of a trick played on your eyes called the Moon Illusion.
Rick Stankiewicz

true for how we see the constellations. Cassiopeia, for example, looks much larger close to the horizon than high up in the sky. As for the moon's rich orange colour, this is due to a physical effect. When we see the moon low in the sky, we are looking at it through a greater amount of atmosphere than when the moon is overhead. When the moon is near the horizon, its light must pass through a lot more atmosphere than when it is high in the sky. We are actually looking through about three times as much atmosphere when the moon is rising or setting. Air molecules and dust in the atmosphere scatter away the blue, green, and purple components of white moonlight (actually reflected sunlight) and thereby allow the longer wavelengths of light like orange, yellow, and red to dominate.

OCTOBER

The Time of Falling Leaves

The native maples of central and eastern Ontario. Note that the mountain and striped maples are shrub-sized, understory trees.

Kim Caldwell

Just before the death of flowers,
And before they are buried in snow,
There comes a festival season
When nature is all aglow.

— Author Unknown

OCTOBER IS USHERED IN BY FLAMING LEAVES OF RED, ORANGE, AND YELLOW — and the conviction most autumns that the colours must be "the most beautiful in years." The early October sun shines with warm benevolence and casts a hazy, surreal light. Crickets sing in meadows of aster and all is gentle and still. Winter seems far away. But, as experience has taught us, the beauty of early October is both temporary and fragile. So we try to hang on to these magnificent days before wind and rain scatter leaves and pre-winter descends upon us. Perhaps it is the ephemeral nature of October's loveliness that makes it so special.

The southward flight of many birds continues this month. Hardier species, such as geese, ducks, and sparrows, are now making their journey. Many of the sparrows, in particular, will linger during their southward passage and become regular visitors to backyard feeders.

With the first heavy frosts, fields, rooftops, and windshields are covered in silver. Leaves drop from trees like a light rain and vistas that were previously hidden by foliage once again become visible. The leaves gather everywhere and can make cycling treacherous, especially on wet days. October, too, is the month of the rake. But there is a payoff for our labour in the familiar, spicy smell of the fallen leaves. It is a smell which seems to transport us back to childhood, evoking an instant flood of memories of autumns past. To truly appreciate this wonderful aroma, however, allow some time for a walk in the autumn woods. At no other season of the year is our sense of smell so fully engaged as when walking through a deep carpet of fallen leaves on a mild, damp October day. Although it is far less common today, burning leaves was always considered a rite of fall, too. Like the smell of pumpkin pie and a turkey cooking in the oven, the aromatic smoke of a leaf fire can communicate the time of the season as accurately as any calendar.

As October draws to a close, the only leaf colour that remains is the yellow of poplars and tamarack and the browns, oranges, and burgundy of the red oaks. Cornfields and cattail marshes have become a sea of dull yellow. The fallen maple leaves have quickly lost their colour and turned a ubiquitous brown. With cold, wet weather and increasingly shorter days, it's not hard to imagine why the Celts chose this time of year to celebrate the various traditions that have become our Halloween.

BIRDS

OCTOBER HIGHLIGHTS

All Month Long
- Flocks of local Giant Canada Geese are a common sight in the more urban parts of central and eastern Ontario as they fly to and from feeding areas such as corn and soybean fields.

- Having completed nesting duties in the Arctic, large numbers of migrating Snow Geese stop over to feed in grain fields in southeastern Ontario. The birds will linger here until winter conditions force them to depart for their wintering grounds along the American mid-Atlantic coast.

- Family groups of Eastern Bluebirds, sometimes numbering over 20 birds, roam the autumn countryside. Flocks of gulls, blackbirds, and robins are still widespread.

- Sparrow migration takes centre stage this month. Watch especially for the beautiful, thrush-like Fox Sparrow feeding on the ground with large numbers of White-throated Sparrows, as well as Dark-eyed Juncos and White-crowned Sparrows. All four species can be attracted to your yard by spreading black oil sunflower seed and niger seed on the ground.

- White-throated Sparrows sometimes break into a short, half-hearted version of their well-known "Oh-Sweet-Canada-Canada-Canada" song. As the late Doug Sadler, a Peterborough area naturalist, observed, there is a "recrudescence of the life-force" that wells up in many birds and frogs in the fall and manifests itself as song, even though its reproductive purpose is for another season.

- On balmy October days, male Ruffed Grouse can sometimes be heard drumming. Young grouse leave the family group at this time to establish their own individual territories. However, their wandering is often accompanied by strange behaviour. October grouse have been known to fly into walls, fences, windows, and just about any other obstacle imaginable.

- Birders keep an eye open in the late summer and fall for marsh birds from the southern United States that sometimes drift northward. The classic example of a post-breeding wanderer is the Cattle Egret. Great Egrets, too, often disperse inland from Georgian Bay and Lake Ontario breeding colonies and can turn up in many parts of central and eastern Ontario. Totally unexpected species from the western states and provinces may also show up, sometimes as a result of far-off storms which can knock birds far off their migration routes.

- Northern Saw-whet Owls migrate southward through central and eastern Ontario from mid to late October. In a research study led by Dr. Erica Nol of Trent University, the owls are caught at night using fine netting and a tape recorder playing Saw-whet Owl calls. Any birds caught are untangled, weighed, measured, and banded. A key finding so far is that these owls are very nomadic. Saw-whets banded by Trent near Bobcaygeon have been re-caught in places as far away as Virginia, Maryland, Wisconsin, and Missouri.[1]

- Nomadic flocks of Purple Finches often pass through central and eastern Ontario and show up at feeders. When food is plentiful, large numbers sometimes overwinter here.

- Die-offs of large numbers of water birds such as loons, ducks, and terns occur from time to time on the Great Lakes in the fall as a result of Type E botulism. The toxin is produced by a bacterium that lives in lake-bottom sediment. Under certain conditions, the toxin enters the food chain and appears to become concentrated in zebra mussels and mussel-eating fish such as Round Gobies. Both of these are non-native, invasive species. Water birds, many of which are fall migrants, are poisoned when they eat the mussels or the fish and often drown as a result of paralysis.[2]

Early October

- Golden-crowned and Ruby-crowned Kinglets, along with Yellow-rumped Warblers, Brown Creepers, White-throated Sparrows, and Dark-eyed Juncos are migrating in large numbers. Listen for the gentle, ridiculously high-pitched calls of the kinglets as they search for food in mostly conifers along roads and trails.

Mid-October

- Rough-legged Hawks that have just finished breeding in the Arctic tundra are passing through. This large hawk frequents open farmland and often hovers as it searches for small rodents. Many will linger here for about a month before continuing south, some no farther than Amherst Island near Kingston. We tend to see mostly light-coloured Rough-legged Hawks in central and eastern Ontario.

- Migrating flocks of diving ducks (e.g., goldeneye, scaup, and mergansers) are congregating on larger inland lakes, the Great Lakes, and the Ottawa River as they move southward. However, fall waterfowl migration is usually less spectacular than in the spring and human use of the lakes and rivers will scare them away. Quite often, sewage lagoons are one of the best places to see ducks during fall migration, especially inland from the Great Lakes.

The White-throated Sparrow is one of the most abundant October migrants.
Karl Egressy

Late October

- The first winter finches often show up. Unfortunately, an influx of these birds isn't always an annual event and is difficult to predict. (See below)

- Mid to late October sees the arrival of Northern Shrikes, migrating south from the Arctic. These robin-sized birds are typically seen perched at the very top of a tree or shrub, making them easy to spot. Shrikes are well known for their habit of impaling prey items such as mice and small birds on thorns or barbed wire. In fact, their scientific name, *Lanius excubitor*, means "butcher watchman." Such macabre behaviour seems strangely appropriate as Halloween approaches.

FORECASTING WINTER FINCH NUMBERS

Each fall, information is compiled on the relative abundance of seed crops in Ontario in order to make educated guesses as to how bird movements may be affected. For many years, this has been done by Ron Pittaway, a former MNR employee.[3] Much of the data comes from Ministry of Natural Resources staff across the province. Movements of Red-breasted Nuthatches, for example, are linked to the cone crop on white spruce, balsam fir, and white pine. When the crop is poor in the north but good farther south, lots of these hyperactive birds are usually present in central and eastern Ontario all winter. Pine Siskins, too, are big fans of conifer seeds, as are Red and White-winged Crossbills. The latter prefer the small, soft cones of white and black spruce as well as those of eastern hemlock. Both crossbill species will fly thousands of kilometres in search of food and will occasionally even turn up in cities.

An abundance of acorns and beechnuts can convince large numbers of Blue Jays to hang around rather than migrating south into the United States. Common Redpolls, on the other hand, get by largely on white and yellow birch seeds in winter. If the birch seed crop in central and eastern Ontario is large enough, and if it fails to the north of us, redpolls will usually put in an appearance. You can then expect at least some of them to show up at your niger seed feeder.

Whether or not Pine Grosbeaks move south depends mostly on how much fruit is on the wild American mountain ash. Central and eastern Ontario residents with a European mountain ash tree in their yard are often paid a visit by flocks of this beautiful finch in years when mountain ash berries are scarce in the north but more abundant farther south. Algonquin Park is usually a good place to see winter finches, both along the highway and at the visitor centre near the east entrance to the park. Most winters, the feeder attracts large numbers of finches, including Evening Grosbeaks. Sometimes pine martens and even fishers come in to dine on the suet and sunflower seeds. Eastern wolves, too, are occasionally seen from the observation deck, feeding on road-killed moose sometimes put out by park staff.

MAMMALS

OCTOBER HIGHLIGHTS

All Month Long

- Coyotes continue to be very vocal this month and throughout the fall. (See page 237)

- Deer mice seek out winter accommodation. Cavities such as a woodpecker hole or even a man-made birdhouse are favourite winter shelters, but human habitations are often chosen as well. Deer mice — and their droppings — carry the Hantavirus. Although rare, the virus is potentially deadly.

- Watch for white-tailed deer feeding along the edges of cornfields and woodlots at dawn and dusk. In much of central and eastern Ontario, we have inadvertently created high deer population levels by carving up forest habitat into small blocks, which increases the amount of grassy edge habitat and provides supermarket conditions for deer. We also farm much of what is in-between the woodland blocks with crops that deer love, especially corn.

- Using its antlers, buck white-tailed deer make scrapes in the leaf litter on the forest floor in preparation for the upcoming rut. The buck then urinates on its hind legs in such a way that the urine runs over special "hock" glands and carries a scent from the glands down to the scrape. Female deer visit these scrapes.

- As in the spring, mole activity on lawns is often evident. As they excavate tunnels just under the ground surface, moles push up low ridges of soil, which often cover large sections of lawns. Volcano-shaped mounds of dirt are common, too. Star-nosed moles are usually the culprits on lawns bordering marshes and swamps. On drier sites, the excavation is usually the work of hairy-tailed moles.

Early October

- Groundhogs and jumping mice start their long period of hibernation. (See page 238)

Late October

- Chipmunks retreat to their secure, well-provisioned dens. Unlike the groundhog, chipmunks are unable to store enough body fat to get them through the winter without eating.

- Porcupines mate from late October through December and employ an elaborate dance. They can look quite comical as they roll and tumble, stand up on their hind legs, and actually box and cuff each other. Fortunately for the male, the underside of the female's tail does not have quills, so he is able to lean against her and mate in typical mammal fashion.

- If you find a Halloween bat in your house, it is probably a big brown, a species that often overwinters in buildings. Little browns, on the other hand, choose caves and abandoned mines as winter quarters.

The female porcupine is only in heat for 8 to 12 hours during the entire year.
J. Glover, Wikimedia

VOCAL COYOTES

Although most people only see coyotes on rare occasions, hearing them is quite common. Coyotes can bark, squeal, growl, and wail, but what we hear most often is a series of high-pitched yips and howls. Dawn and dusk tend to be the peak calling times; however, the animals can be heard at other times of the day and night, as well. When a coyote begins to call, a second animal often joins in. When the two sing in unison it can create the illusion of a dozen or more. The howling and yipping tend to start very suddenly and last for only 30 seconds to two minutes. It then usually stops as abruptly as it started. Why coyotes are so vocal is a bit of a mystery. It may have something to do with territorial claims, a cry for a mate, or possibly just an expression of joy or sociability. Just like your dog, coyotes are very social animals and love to communicate with one another.

Although we tend to hear coyotes most in the spring and summer — probably because our windows are open — the peak calling times are actually August through October and then again in mid-winter. In the fall, the pups are well into their adolescence and seem to enjoy experimenting with their newfound voices. They have a higher-pitched howl than the adults and also make bird-like squeaks and squeals. Late January to March is the mating season and another time when coyotes need to communicate. For anyone wanting to go out and listen for coyotes, try driving through areas of marginal farmland just after dark or at dawn. Stop from time to time and listen. If all is quiet, try howling yourself. By creating a high-pitched, mid-volume imitation of the classic coyote howl (*yip-yip-yerrrr*), you may very well elicit a response.

OUR TRUE HIBERNATORS: GROUNDHOGS, BATS, AND JUMPING MICE

Sometime in late September or the first half of October, groundhogs begin their five-month hibernation period. The winter den is often built along the forest edge or even in the forest itself. Being one and a half to two metres in depth, it is well below the frost line. Unlike most other members of the squirrel family, the groundhog does not store food for winter consumption but relies on a substantial layer of fat to provide nourishment during hibernation. The groundhog's heart rate falls from 100 beats per minute to just 15 and its body temperature will eventually drop from 35°C to 6°C!

Bats that overwinter in Ontario are also able to drastically lower their body temperature during winter hibernation. By allowing their temperature to fall all the way to 4°C, they are able to save a great deal of energy. This is important because bats cannot feed during the winter and must therefore depend on their fat reserves. Unnecessary disturbances can cause bats to awaken and use up their energy reserves and therefore not survive to spring. White-nose syndrome is now a major threat to overwintering bats. (See January Mammals)

The only other true hibernators that we have in central and eastern Ontario are the woodland jumping mouse and meadow jumping mouse. They retreat to their winter quarters at about the same time as the groundhog. Jumping mice dig a special winter hole of about one metre in depth, where they then prepare a bed of dried vegetation. To ensure complete privacy, the animal then plugs the entrance to the hole. They are very profound hibernators and will not be seen again until May.

AMPHIBIANS AND REPTILES

OCTOBER HIGHLIGHTS

All Month Long
- The combination of bright, sunny days and cooler air temperatures often compels reptiles such as eastern garter snakes to soak up the available heat by basking on rocks and road surfaces. They are the last of the snakes to retreat to winter quarters.

- Frogs and toads are sometimes found sitting motionless on warm road pavement on wet, mild fall nights. The temperature difference between the cooler ambient air and the heat-retentive road surface probably explains their popularity with amphibians.

- October is a great time to find salamanders. Red-backed salamanders, which are almost worm-like in appearance, are especially easy to find. Look under flat rocks, old boards, or logs in damp wooded areas and around cottages. Be careful, however, to gently put the rock, log, or board back in its original position. In your searching, you may also encounter both the spotted and blue-spotted salamanders.

- Sometime in late October or early November, massasauga rattlesnakes enter their hibernaculum — often in rock fissures or animal burrows — and will re-emerge only with the consistently warm days of April.

- Turtles are seen only rarely after mid-October. The shorter days and weak fall sun do not provide the necessary warmth to raise their body temperature sufficiently to allow them to catch and digest food. Consequently, many turtles begin hibernation. The last species seen is often the map turtle.

FISH

OCTOBER HIGHLIGHTS

All Month Long

Brook Trout can only spawn in areas where there is an underwater spring with water coming up through the sand or gravel.

Ministry of Natural Resources

239

- With the "autumn turnover" (see Weather) and cooler water temperatures, some species such as bass are becoming less active.

- Muskellunge continue to feed heavily. This can make for great fishing.

Early October

- Lake Trout are spawning along shorelines with wind-exposed areas of rock rubble. Spawning activity takes place only at night, when water temperatures are between 10°C and 12°C. The depth of the water can range anywhere from 30 centimetres to 20 metres. No attempt is made by the adults to guard the eggs.

Late October

- Brightly coloured Brook Trout provide one of the last colour spectacles of the year as they spawn at gravel-bottomed sites in stream headwaters and along lakeshores. (See below)

COLOUR TO RIVAL THE FALL LEAVES

As the last of the fall colour drains from the landscape and distant hills turn grey, a lesser-known colour spectacle is gearing up right now in some of our lakes and streams. Unlike the display of the leaves, the colour show put on by spawning Brook Trout is one that few of us ever see. How unfortunate that is, because Brook Trout right now are offering one of nature's last colour spectacles of the year.

In late September, male Brook Trout, or speckled trout if you prefer, begin to acquire a deep orange to red underside that distinguishes them during the spawning season. An elegant black border on the underside separates the white belly from the rich flank colour. The orange and red pigments are acquired from the bodies of certain crustaceans on which the trout prey. During the summer these pigments are stored in the muscles, giving the flesh its distinctive pink colour. In the fall, these same pigments move from the muscle to the skin in males. As in birds, the colours may serve to make the male fish more attractive to the female, as well as being an indication of overall health and suitability as a mate.

Spawning itself will take place any time from late October to late November. Brook Trout usually choose gravel-bottomed sites in the shallow headwaters of streams but will also lay their eggs along lake shorelines. However, there must be an upwelling of spring water under the spawning bed. The spring water, which is warmed by geothermal heat from deep within the Earth, keeps the eggs at a warmer temperature during the winter months and allows for successful egg development. If you are fortunate enough to know of a location where Brook Trout spawn, you may be able to observe firsthand the mating spectacle. Caught up in the urge to reproduce, the fish often seem oblivious to human presence. As they mill about in the shallow water, you should be able to catch glimpses of the dazzling red-orange flanks of the male, as every once in a while he will turn and catch the sunlight at just the right angle. It's ironic that just as the natural world seems to be shutting down for winter, both of our native trout species are busy renewing life with the promise of a new generation. Some of the best places to see spawning Brook Trout are in Algonquin Park in lakes such as Dickson.

INSECTS AND OTHER INVERTEBRATES

OCTOBER HIGHLIGHTS

All Month Long

- Sulphur and mustard white butterflies are active throughout October. The occasional monarch and mourning cloak can also be seen.

- If you come across a mosquito this month, don't be overly surprised. Mosquitoes of the genus *Culex* are still somewhat active, albeit in small numbers. They overwinter as adults in subterranean structures such as basements and sewers.

- Small mating clouds of midges are still common, even on days when the temperature is just above freezing. It is amazing that these delicate insects are able to fly at such low temperatures when much larger and more robust insects such as butterflies are unable to coax their wing muscles into action.

- Some darners, as well as small red (male) or yellow (female or immature male) meadowhawk dragonflies are still active.

- On warm days in mid-fall, watch for strands of spider silk floating through the air or caught up in branches. As explained in May, a baby spider is attached to each strand, or "gossamer." Although this "ballooning" also occurs at other times of year, it seems to be most noticeable in October. Hence, this special period of the year is sometimes called the gossamer days.

- Cluster flies are drawn to human habitations in the fall. If you happen to have a healthy lawn with abundant earthworms — the larvae parasitize worms as part of their life cycle — you may see hundreds of these flies "clustering" on sunny walls. Slightly larger than a housefly, these bristly grey insects have numerous fine, yellowish hairs on their back. They hibernate by jamming themselves in large numbers into roof and wall cavities of buildings. On mild, sunny winter days, the flies often wake up and may succeed in getting into the house. They then congregate on windowpanes and buzz frantically.

Mid-October

- Killing frosts usually bring the cricket and grasshopper chorus to an end and terminate most insect activity in general.

- Wasps, flies, and ladybird beetles gather on the sun-warmed sides of buildings in an effort to warm their flight muscles. It is not uncommon to see birds such as migrating Yellow-rumped Warblers gobbling them up.

PLANTS AND FUNGI

OCTOBER HIGHLIGHTS

All Month Long

- Flowers continue to go to seed and to use many ingenious strategies for dispersal from the parent plant. (See page 244)

- Trees are shedding their leaves, each species following its own timetable. Even when there is no wind, the warming effect of the morning sun is sufficient to cause many leaves to fall. (See page 245)

- The whites, purples, and mauves of asters completely dominate fields and roadsides.

- The first ground frosts usually mean that the mushroom-hunting season is nearing its end. However, a couple of genera are stimulated to grow in the cold weather. They include some of the waxcaps (*Hygrophorus*) and oyster mushrooms (*Pleurotus*).

- Looking very much like old porcelain doll's eyes, each with its own prominent black dot, the china-white fruits of the white baneberry stand out conspicuously on their thick red stems. Watch for them close to the forest floor in rich soils. Please also note that the berries are poisonous.

Early October

- The fall colours are at their height and quite often coincide with Thanksgiving weekend. The sugar and red maples put on the best colour show. The spectacle doesn't last long, however, especially if there is wind and rain.

- When the leaves change colour, the identity of individual trees is often instantly apparent, even from a distance. Trees that were simply part of the "green blur" all summer suddenly stand out in their own unique fall colour, boldly advertizing what species they are.

Mid-October

- As the oranges and reds of the maples fall to the ground, the yellows take over. Most of the colour is being supplied courtesy of trembling aspen, bigtooth aspen, balsam poplar, silver maple, and white birch. Interspersed here and there are the reds, browns, and burgundies of the oaks.

- Although we spend a great deal of time raking, the rich, spicy fragrance of the leaves and the cool, invigorating air make the effort worthwhile. Vistas previously hidden by the leaves now become visible again.

Late October

- You can get a real sense right now of just how many of our city trees and shrubs are non-native. They remain stubbornly green and appear oblivious to the colour change occurring all around. (See page 245)

- The last of the asters bring the year's wildflower parade to a conclusion.

- The smoky, golden-yellow of tamaracks lights up wetland borders. All summer long, they went unnoticed, but they now stand out like yellow beacons. Glowing tamaracks also represent the final act in the annual fall tree colour extravaganza.

Asters, such as this heart-leaved aster, provide the last of the year's offering of pollen and nectar to wasps and bees.
Drew Monkman

HAVE GENES, WILL TRAVEL

Flowers exist for only one reason: to create a new generation. The final step in the process is the dispersal of the flowers' seeds to sites that will be suitable for germination and growth. Seeds must get away from below the parent plant if they are to survive. Try to imagine the insurmountable difficulty a tiny seedling would face if it had to compete with the parent plant towering above it for sunlight, nutrients, water, and space. Since adult plants have no choice but to remain rooted where they are, their embryos must do the moving. The special adaptations that seeds have evolved for travel are fascinating. Some seeds are sticky or bristly and are able to hitch a ride on a mammal's fur or a walker's pants. Beggar's tick (barbs), tick-trefoil (hooks), and burdock (Velcro-like hooks) all use this method. Other seeds, such as those of cherries, blueberries, and dogwoods, wrap themselves in a bright, tasty fruit in order to move by way of a bird's, bear's, or raccoon's digestive system. Although the digestive process breaks down the nutritious outer coating of the fruit, the seed within usually passes through the animal's gastrointestinal tract unharmed. An added bonus is that the seed usually ends up being deposited right in the middle of a pile of mineral-rich scat. As for appearance and nutritional value, bird-dispersed fruits are usually red or black and rich in energy in the form of sugars. So, from an evolutionary perspective, the decisions of countless generations of birds — animals with keen eyesight — have forced plants to evolve conspicuously coloured, energy-rich, fleshy fruits to attract the birds that will disperse them. However, species like oak, beech, and walnut, which have coevolved mostly with mammals, some of which have poor eyesight and rely more on odours, tend to have dull-coloured but smelly fruits.

Wind dispersal is another common strategy. Many plants have evolved an elevated umbrella-like crown of intricately branched hairs at the top of the seed. Acting like a parachute or sail, even a light gust of wind will often catch the hairs and raise and propel the seed into the air. Milkweed seeds are dispersed in this manner and typically drift 30 metres or more before settling to earth. As for trees, poplars, aspens, and willows have also evolved this adaptation to seed dispersal. A widespread alternative to packing a parachute can be found in seeds that possess a rigid or membranous wing-like appendage. Common examples are the single twirling blade of the ash and basswood and the double helicopter-like rotors of the maple. Like a propeller, the seed's blade typically has an asymmetrical profile which forces the air to flow faster over one side than the other. This causes the seed to spin in helicopter-like fashion as it falls. At the same time, a small amount of lift is created, which can carry the seed considerable distances. Most evergreen seeds are also wind travellers but simply rely on a wing-like margin that enables them to glide as they fall. Finally, a small number of plants disperse their seeds themselves by having evolved active shooting devices. In different ways, plants build up tension internally, which provides the force to shoot the seeds impressive distances. A common example is jewelweed, also known as touch-me-knot. As Scott Camazine writes in *The Naturalist's Year*, "There is no master plan, no grand design, that has been followed in the construction of these seed dispersal adaptations. Here is a lesson to be learned about all of nature, not just the means by which seeds are scattered."[4]

The seeds of the common milkweed peel away from the pod a few at a time and are carried off by the wind.
Drew Monkman

THE LEAF DROP TIMETABLE

The following chart shows the approximate time when the more common trees, shrubs, and vines have shed most of their leaves.

Early October: White ash, red maple, staghorn sumac, Virginia creeper.

Mid-October: Sugar maple, black walnut, American basswood, American elm, pin cherry, red osier dogwood.

Late October: White birch, trembling aspen, bitternut hickory, American beech (mature trees), speckled alder, silver maple.

Early November: Tamarack, Norway maple, red oak, European buckthorn, common lilac, black locust, native willows.

Mid-November: Siberian elm, weeping willow.

WHY ARE SOME TREES STILL GREEN?

As the month advances, it quickly becomes apparent that many trees in the city and around country homes are still completely green and seemingly unaware that fall has arrived. These trees and shrubs are almost all non-native species or "exotics," and they are still operating on their ancestral Eurasian timetable. Some of the common species include Norway maple, King Crimson maple (a variety of the Norway with burgundy leaves), weeping willow, little-leaved linden, Siberian elm, Carolina poplar, common lilac, tartarian honeysuckle, and European buckthorn. The buckthorn in particular has become very widespread and is now the most common shrub (or small tree) in much of urban and suburban central and eastern Ontario, in the understory of neighbouring woodlots and along hedgerows bordering on agricultural lands. Its abundance is immediately apparent at this time of year because it is still green when just about everything else is leafless. Its fruit is quite popular with robins and waxwings. Unfortunately, however, the berries are also a strong laxative, which results in purple bird dropping stains on concrete and patio stones. Although some people may feel that non-native species allow us to enjoy the greenery of summer just a little longer, they also tend to stick out like a sore thumb at this time of year. Because these same trees have been planted throughout much of North America, they take away from the sense of place that native species provide.

WEATHER

OCTOBER HIGHLIGHTS

All Month Long

- We lose another 90 minutes of daylight this month. We begin with 11 ¾ hours but end the month with only about 10 ¼.

- Generally warm weather is still with us for most of October. As a matter of fact, an extended period of days above 20°C occurs most years. When this warm weather comes after the first "hard frost" — a frost where both the air and the soil have dropped below freezing — it is referred to as Indian Summer.

- The spicy fragrance of fallen leaves is October's signature smell. For many, it is associated with feelings of nostalgia. Although far less common now, the smell of burning leaves, too, takes many people back to the days of their childhood.

- On warm, sunny October days, the light often has a hazy, almost dream-like quality. This is due to the large amount of water vapour in the air in the fall and the fact that the sun is lower in the sky. These two factors combine to produce a feeling of reverie and stillness.

- With the long, cool nights of fall, the surface waters of our lakes eventually cool down to the same temperature (4°C) as the uniformly cold, deeper waters below. All of the water now has the same resistance to currents and no longer does the lake function as two separate lakes, with a warm lake on top of a cold one. The currents, which are caused by wave action, can therefore thoroughly mix and oxygenate the entire lake, just as in the spring. The "fall turnover" will continue until freeze-up. At times, the mixing even brings dead weeds and other debris from the lake bottom to the surface.

- This is usually the month of the first killing frosts and the first strenuous scraping of the car windshield.

- In the fall and winter months, release of the heat stored in the Great Lakes moderates the temperature near the shores of the lakes. This "lake effect" is especially evident in the Bruce Peninsula, where some plants bloom late into the fall.

Late October

- The first snowfall usually whitens the ground for a brief period by month's end.

- Total darkness is upon us by just after 6:30 p.m.

OCTOBER WEATHER AVERAGES, 1971–2000[5]

City or Town	Daily Max. (°C)	Daily Min. (°C)	Rainfall (mm)	Snowfall (cm)	Precipitation (mm)
Owen Sound	13.3	5.7	86.5	2.3	88.8
Huntsville	12	3.1	94.7	3.1	97.8
Barrie	13.2	3.7	74.3	2.5	76.8
Haliburton	11.7	2.2	84.3	5.2	89.4
Peterborough	13.3	3.1	68.4	2.3	70.6
Kingston	13	4.4	86.4	1.1	87.5
Ottawa	12.6	3.7	76.2	3	79.1

APPROXIMATE OCTOBER SUNRISE AND SUNSET TIMES (DST)[6]

(Note: Twilight starts about 30 minutes before sunrise and continues about 30 minutes after sunset.)

Location	Date	Sunrise (a.m.)	Sunset (p.m.)
Grey-Bruce	Oct. 1	7:21	7:07
	Oct. 15	7:38	6:41
	Oct. 31	7:59	6:16
Muskoka/Haliburton	Oct. 1	7:16	7:02
	Oct. 15	7:33	6:37
	Oct. 31	7:54	6:11
Kawartha Lakes	Oct. 1	7:10	6:56
	Oct. 15	7:27	6:31
	Oct. 31	7:48	6:06

Kingston/Ottawa	Oct. 1	7:03	6:49
	Oct. 15	7:20	6:24
	Oct. 31	7:41	5:59

NIGHT SKY

OCTOBER HIGHLIGHTS

All Month Long

- Major constellations and stars visible (October 15 — 9:00 p.m. DST)
  Northwest: Ursa Major (low in north) with Ursa Minor (with *Polaris*) above; Summer Triangle made up of *Vega* (in Lyra), *Deneb* (in Cygnus), and *Altair* (in Aquila) high in the west.
  Northeast: Cassiopeia with Andromeda to right and Pleiades to lower right.
  Southeast: Pegasus (with the Great Square) and Andromeda (with M31 galaxy).
  Southwest: Sagittarius at west horizon; *Altair* (in Aquila) above and to left; Milky Way runs up from the horizon through Sagittarius and Aquila.

- The Orion constellation looms in the southern sky as we head off for work in the early morning darkness. Like the falling leaves, Orion's arrival is a sure sign that winter is fast approaching.

- At any given time of year, at least one of the planets is visible in the night sky. The planet parade is normally made up of Venus, Jupiter, Saturn, Mars, and sometimes Mercury. Because there are relatively fewer bright stars in the fall, any planets that may be visible tend to stand out more now than in the other seasons. However, the comings and goings of planets is not linked to any particular season. In other words, a given planet does not necessarily appear at a given time of year.

- The brightest and most striking planet is Venus. Never very high up, Venus appears on the eastern horizon as the "morning star" or on the western horizon as the "evening star." It is much brighter than any of the true stars.

- The Algonquian name for the full moon of October is the Moon of Falling Leaves or the Hunter's Moon. Like the Harvest Moon, it rises only about 30 minutes later from one night to the next. This means that when the moon is full, or nearly full, there is no long period of darkness between sunset and moonrise. This allowed Native American hunters to pursue game late into the evening by autumn moonlight.

November

A Hush upon the Land

The bare-limbed hardwoods of the quiet November woods.
Kim Caldwell

It is a joy to walk in the bare woods.
The moonlight is not broken by the heavy leaves.
The leaves are down, and touching the soaked earth,
Giving off the odors that partridges love.

— Robert Bly, *Solitude Late at Night in the Wood*

In November a hush settles upon the land. The soft contact calls of migrant kinglets and White-throated Sparrows fade away, most crows and robins depart, and the last crickets surrender to the cold. Damp, cloudy weather, leafless trees, and faded grasses and flowers create a world of mostly greys and browns, punctuated only by the dark green of conifers. Yet sometimes late fall's typical bleakness is pushed aside by a lingering Indian Summer that gently eases us into winter. In other years, the snow comes early and stays until spring.

Like the first Red-winged Blackbirds in March, the arrival of the birds of winter serves to mark the change of season. In addition to the Northern Shrikes and American Tree Sparrows that began arriving in October, other northern visitors such as siskins and redpolls often grace us with their presence. At the same time, however, loons are departing for the Atlantic seaboard and taking with them the last vestiges of summer. For lakeside residents, it is a melancholy event.

A walk on a late November day can seem uneventful, with seemingly little of interest to catch our attention. Yet the relative scarcity of plants and animals allows us to focus on the commonplace — the leafless trees reduced to their elemental form, the intricacy and diversity of the mosses and evergreen ferns, and the beauty of a milkweed pod spilling its last seeds. But, other than the occasional call of a chickadee or woodpecker and the steady rustling of squirrels and mice as they forage for seeds, the woods are nearly devoid of animal sounds. With colder weather, nature's kaleidoscope of smells is also reduced to a minimum. Apart from the scent of decaying leaves or the smoke of a wood stove, there is little to stir our sense of smell. But the cold of late fall does bring renewed appreciation for the warmth and comfort of our homes and anticipation for the holiday season just around the corner. With the yard work done and wood stacked in the garage, we can sit back and enjoy the sound of the north wind as it ushers in winter.

BIRDS

NOVEMBER HIGHLIGHTS

All Month Long

- Small numbers of tiny, beautifully marked Golden-crowned Kinglets are still present. Theirs is an amazing story of evolution. (See page 252)

- Red-tailed Hawks, mainly from the boreal forest of northern Canada, migrate south into the United States. On a good day, hundreds can be seen flying along the north shore of Lake Ontario. Many central and eastern Ontario Red-tails are resident birds, however, and don't migrate. This species is a common sight along Highway 401 all winter long. Roadside grasses make for excellent mice and vole habitat and therefore a perfect place for hawks to hunt.

- Snowy Owls may begin to arrive. This species can be common on the Bruce Peninsula and on Amherst Island in late fall and winter. Winter Snowy Owls are often tied to water bodies, since waterfowl can constitute a large part of their diet.

- As long as there is open water, diving ducks such as mergansers and goldeneye, along with small numbers of Common Loons, will continue to linger on inland lakes and rivers. Diving ducks are also migrating southward along the Lake Huron shoreline. Some days their passage is almost non-stop.

- Feeder activity tends to slow down as migrant sparrows have now left. With luck, however, northern finches will fill some of the void. If you have a crabapple or mountain ash tree, you may receive a visit from a flock of Pine Grosbeaks or Bohemian Waxwings. The latter appear almost every year in at least some parts of central and eastern Ontario.

- Ring-billed Gulls are one of the most commonly seen birds in November. They spend the nights on large lakes where they are safe from predators. With freeze-up, most will head to the Gulf States.

- In years where there is an abundant crop of acorns, you can expect to see large numbers of Blue Jays throughout the fall and winter. Sufficient food is usually enough to convince many of these semi-migratory birds to stick around instead of heading down to the central and southern United States.

Early November
- Most of our loons and robins head south. However, a small number of robins regularly overwinter in central and eastern Ontario, especially in years when wild food is plentiful. (See page 252)

Mid-November
- Glaucous and Iceland Gulls, two Arctic-nesting species, sometimes show up on the larger lakes and at landfills. Look for both species among flocks of Ring-billed, Great Black-backed, and Herring Gulls.

- Although most of our crows depart around this time, a significant number do spend the winter in central and eastern Ontario, especially in areas south of the Shield. Crows winter in huge numbers in southwestern Ontario and in the border states.

Late November
- Male Great Horned Owls stake out breeding territories and become quite vocal. A useful mnemonic to remember their call is "Who's awake? Me, too."

- Usually beginning in late November, resident eagles that nest in parts of central and eastern Ontario are joined by migrants, most likely from northern and northwestern Ontario. Primarily scavengers, eagles are attracted to areas such as the Kawartha Lakes in part because of the high deer population and the relative abundance of carcasses. Eagles can also be found along open watercourses where they often scavenge with gulls. Many of the eagles seen are dark-plumaged immatures, which lack the white head and tail of the adult.

ROBINS AND LOONS DEPART

One of our most familiar summer birds, the American Robin, leaves central and eastern Ontario in late October or early November for the southern United States. Every year, however, some robins try their luck at overwintering here, especially when the fruits of European buckthorn, riverbank grape, European mountain ash, and ornamental crabapples are abundant. Robins are especially fond of buckthorn berries after the berries have had a chance to freeze for a period of time. Freezing sweetens the berries and makes them almost the equivalent of an avian ice wine.

Most loons also leave in early November, although some will have departed earlier. A few will even stay into December, provided there is open water. Dressed in their grey and white winter plumage, Common Loons could almost be mistaken for a different species at this time of year. Immature birds acquire this plumage during the summer, while adults moult to their winter garb in the fall. Adult loons tend to leave central and eastern Ontario before their young. When loons are occasionally trapped by early ice formation, they are usually young-of-the-year birds and probably late-hatching individuals not yet capable of sustained flight. Iced-in loons are sometimes preyed upon by eagles.

THE KINGLET'S LESSON OF EVOLUTION

In late fall and throughout the winter, small numbers of tiny, beautifully marked Golden-crowned Kinglets can often be found in thickets of coniferous trees such as spruce and fir. It is remarkable, however, that such a diminutive species as a kinglet is even able to survive here, since you would think that the fat reserves of such a small bird would be insufficient to get it through a long, frigid winter night. Clearly, those individuals that do overwinter here never stop from dawn to dusk in their search for insect food. Studies of kinglet stomach contents in winter have shown that they subsist on tiny frozen caterpillars. They also huddle at night in groups of two or three in order to save energy.

Still, there is a high level of winter mortality. In fact, it is believed that an amazing 87 percent of the kinglet population is weeded out every year from one cause or another.[1] Bernd Heinrich, in *Winter World*, calls the kinglet as close to an annual bird (in analogy to annual plants that regenerate each year only by seeds) as any bird gets. However, this incredibly high mortality rate is compensated by the kinglet's nesting behaviour. Whereas the majority of songbirds have four to five eggs per clutch, kinglets lay an amazing eight to eleven. In fact, the eggs actually sit in the nest in two layers — one layer on top of the other! And it's not as if kinglets are satisfied with just one nest. They are simultaneously busy with a second nest, which contains the same number of young. This ability to quickly rebuild its numbers each year is clearly the kinglet's saving grace and, in this case at least, an example of how evolution sometimes cares little about the survival of the individual but everything about the survival of the species.

To see late fall and winter kinglets, begin to pish when you hear chickadees calling in suitable habitat or when you hear the faint, high-pitched *tsees* calls of the kinglets themselves. They are much shyer than the chickadees, however, and may take a while to come into the open.

Golden-crowned Kinglets are almost always found in small groups of three to six.
Karl Egressy

MAMMALS

NOVEMBER HIGHLIGHTS

All Month Long
- Striped skunks, raccoons, and black bears retreat to their winter quarters but will come out on warm days. They are not true hibernators. (See page 255)

- Muskrats build dome-shaped homes and feeding platforms of cattails and mud. They are only about one metre in height, making them much smaller than beaver lodges.

- A late fall visit to a local wetland can often provide a front-row seat to the secret lives of beavers. Beavers are busy now cutting down large numbers of trees to gather food branches for their winter caches. These branches can be seen sticking out of the water beside the lodge. A pile of fresh branches near the lodge is a sure sign of an active beaver colony.

- Being the hardiest of Ontario's bats, big browns can still be active this month and maybe even seen feasting on the moths of Indian Summer.

- Early snowfalls reveal the nocturnal world of mammal movements. Coyotes, deer, squirrels, mice, and voles are just a few of the many species that leave their tracks for us to decipher.

- The barely audible, high-pitched squeaking and peeping of shrews can sometimes be heard in woodlands on calm days. Listen for soft chirps and squeaks, not unlike those of baby birds softly begging in the nest. Some research has suggested that shrews might use these sounds for echolocation.

- Snowshoe hares and weasels are acquiring their white winter coats. In the case of the hare, the ears and feet turn white first while the back is the last section of the body to change colour. Except for the black ear tips, snowshoe hares are usually pure white by early December. Hares rely on their camouflage to avoid predation; however, with climate change, their future is uncertain. With potentially less snow in the spring and fall, they may be white against a background of brown and therefore stand out conspicuously.

- Now that most of the leaves are down, basketball-sized "leaf-balls" are very evident near the tops of large trees such as maples. These are actually gray squirrel nests or "dreys" and are made of an outer shell of leaves and twigs. A cozy inner chamber is lined with mosses, grasses, shredded bark, and sometimes even cloth or paper. Dreys are most commonly seen in cities. Gray squirrels will also nest in tree cavities or even in the attic or exterior wall of a house.

- Peak breeding season for white-tailed deer in central and eastern Ontario is the second and third weeks of November. The buck's antlers have matured and hardened by this time and the animals are "in rut" — at the peak of their sexual readiness. The dominant male in a given area may mate with as many as seven does.

- More collisions involving deer take place in late October and November than in any other month. When driving after dark, watch for dark shadows along the side of the road and the bright green reflection of the deer's eyes in your headlights. If you do see a deer on or near the road, slow down immediately because there will likely be more deer nearby. They become disoriented and unpredictable when confronted with an automobile. Most accidents involving deer occur at dusk.

Early November

- In much of central and eastern Ontario, the first Monday in November marks the beginning of the annual deer harvest by rifle and shotgun. In some areas, there is still a veritable "flight to the woods," with businesses shut down and most able-bodied men between 16 and 90 away on the hunt.

Squirrel dreys are easy to spot in most urban areas, once the leaves fall.
Drew Monkman

THE RETREAT TO WINTER QUARTERS CONTINUES

As the weather becomes colder and food disappears, many mammals retire to sheltered dens, where they spend most of their time sleeping. Unlike true hibernators, which enter a death-like state of extremely low body temperature and heart rate, mammals that simply sleep for extended periods maintain nearly normal body temperature and awaken frequently.

Chipmunks sleep two or three days, awaken to visit a food chamber to eat, go to another chamber to get rid of body waste, and then return to sleep. Raccoons and skunks do not store food but rely on body fat to get through the winter. When temperatures are below freezing, both of these species sleep for extended periods. However, they often become active again during mild spells. For their winter quarters, raccoons will choose abandoned burrows, caves, hollow logs, culverts, and buildings. Skunks usually prefer a chamber in the ground such as an old groundhog burrow. They will take refuge under buildings, as well. Skunks are generally more active than raccoons in the winter.

Black bears stay close to their denning site in November and will retire with the onset of cold weather. They usually dig out their own den on the side of a hill or under an uprooted tree. Sometimes, however, the den consists of nothing more than the shelter of a brushpile or a rock crevice. In fact, about one-third of adult male bears sleep directly on the ground, often on a pile of bunched up grasses or conifer branches.

AMPHIBIANS AND REPTILES

NOVEMBER HIGHLIGHTS

All Month Long
- Most amphibians and reptiles have already begun hibernating by early November.

- Amphibians and reptiles use a wide variety of strategies to get through the winter. These range from becoming frozen "frogsicles" in the leaf litter to retreating deep into the ground below the frost line. (See below)

FROGSICLES AND OTHER OVERWINTERING STRATEGIES

By November, most reptiles and amphibians have moved to their wintering quarters and will use a number of different strategies to survive the onslaught of winter. Surprisingly, many frog species do not burrow into the mud at the bottom of wetlands but actually spend the winter in the leaf litter of the forest floor. This group includes the spring peeper, chorus frog, gray tree frog, and wood frog. The onset of cold weather is their cue to burrow down several centimetres into the damp leaf litter. As outside temperatures drop, the frog's metabolism slows to a crawl and its body temperature approaches 0°C. Just like the leaves and moisture around it, the frog essentially freezes. However, when the first ice crystals begin to form on the frog's skin, an alarm reaction is set off. Adrenaline is released into the frog's bloodstream which activates enzymes that convert glycogen in the animal's liver to glucose. Blood and cellular glucose levels rocket to astronomic concentrations — levels that would kill a human many times over. Like the antifreeze in your car, the glucose acts to lower the freezing point of the cellular fluid. This protects the integrity of the cell. Remember that when water freezes, it also expands, so freezing within a cell itself would tear it apart.

At the same time as glucose and urea levels are soaring, much of the water within the cells is actually being withdrawn through osmosis. The water accumulates in the frog's abdominal cavity and in the area just under the skin. Here, special proteins and bacteria actually promote the formation of ice crystals. As much as 65 percent of the water in the frog's body gradually crystallizes. However, ice formation in these open spaces is safe and does not interfere with any of the vital organs. Within only 15 hours, the frog essentially becomes a block of ice. Its eyes actually turn white because the lenses freeze! Laboratory studies have shown that wood frogs, for example, can survive minimum body temperatures of as low as -6°C. During the many months of suspended animation, there is no breathing, blood circulation, or heartbeat. All of the cells are therefore deprived of oxygen. By most definitions, the frog is essentially dead. However, when researchers bring these "frogsicles" inside from the cold and let them thaw out, the frogs become active again quite quickly.

Some frog species actually do overwinter in the mud at the bottom of wetlands and other water bodies. These include green frogs, bullfrogs, and mink frogs. They are able to take the little oxygen they need directly through their skin. Leopard frogs prefer moving water and will often migrate to streams and rivers in the fall. The fast-flowing water provides plenty of oxygen.

The American toad has a completely different strategy for winter survival. The toad retreats to below the frost line, either by taking up winter residence in ready-made burrows or crevices or by burrowing down into loose soil. It stops digging when the soil temperature remains at 1°C to 2°C above freezing.

Terrestrial salamanders such as the red-backed salamander use strategies similar to the toad. Aquatic salamanders such as the mudpuppy, however, remain semi-active during the winter and are occasionally seen swimming under the ice. Snakes must also descend below the frost line in order to survive the winter. Rock piles, crevices, and rodent burrows are all used as hibernacula. It is not uncommon to find a number of different snake species using the same wintering site. This is probably explained by the scarcity of good sites.

Aquatic turtles use a strategy similar to many frogs to survive winter, sinking down into the mud at the bottom of small lakes and wetlands. By extending their head and legs in an effort to expose as much skin as possible, hibernating turtles are able to take up dissolved oxygen from the water. Their physical lethargy and low body temperature reduces their resting metabolism to a point where their heartbeat can slow to less than 1 percent of the summer rate. This allows the animals to survive for extended periods of time, even when almost all of the oxygen in the water has been used up.

As mentioned in August, some baby painted turtles actually overwinter in the nest and do not burrow out until the following spring. Like wood frogs, only body fluids outside of the cells freeze, since a natural antifreeze produced by the liver prevents the cells themselves from freezing and becoming damaged.[2] When the warmer temperatures of spring arrive, the turtle thaws out — seemingly unaffected by having been frozen — burrows out of the ground, and heads for the water. When the second winter rolls around, the young turtles must join the adults and hibernate in the bottom of their frozen pond.

So, never think of late fall and winter as a boring season where nothing of interest is taking place. Amphibian dormancy and hibernation are examples of the true miracles of life on this planet.

The wood frog's attractive dark mask has earned it the nickname of "robber frog."
Ontley, Wikimedia

FISH

NOVEMBER HIGHLIGHTS

Early November

- Lake Whitefish are the last species of the year to spawn. They wait until water temperature cools to below 7°C. Spawning continues through early December. The young fish hatch out in April or May.

- Many anglers contend that early November offers the best Muskie fishing of the year.

Late November

- Walleye begin to move upstream along large rivers. They remain in the rivers over the course of the winter in anticipation of the early spring spawn.

INSECTS AND OTHER INVERTEBRATES

NOVEMBER HIGHLIGHTS

All Month Long

- Ball-like swellings known as galls are easy to see on the stems of goldenrod plants. If you open the gall with a knife, you will find the small white larva of the goldenrod gall fly inside. (See below)

- Monarch butterflies are now arriving on their wintering grounds in tiny patches of Oyamel fir forest, high up in the Sierra Madre Mountains of the state of Michoacán, west of Mexico City. According to a study led by Steven Reppert of the University of Massachusetts, they can thank their antennae for having steered them in the right direction. The monarch's antennae are more or less the equivalent of a global positioning system. When the researchers painted the butterfly antennae black, the insects got lost.[3]

- Other than the occasional sulphur butterfly, cluster fly, or meadowhawk dragonfly, few insects are active. Most have already begun diapause, the state of halted development in which insects overwinter. (See page 260)

- A few hardy field crickets may still be heard on warm days.

Early November

- The last dragonflies and butterflies are seen. Some species that may still be active, albeit in very small numbers, include the yellow-legged meadowhawk dragonfly and the clouded sulphur butterfly.

GOLDENROD GALLS

Most people are familiar with the ball-like swellings on the stems of goldenrod plants. These are actually caused by a common roadside fly, the goldenrod gall fly. In early summer, the female fly lays her eggs on the stems of developing goldenrod plants. The eggs hatch and the larvae burrow into the stem and create a chamber in which to feed. The plant responds to this intrusion by growing a spherical deformation around the insect chamber. If you open the gall with a pen knife, you will find a small white larva with a dark head. The larva spends the winter in this cozy enclosure. In the spring, the larva becomes active again and chews out an escape route, almost to the outer surface of the gall. It then moves back toward the centre and pupates in a hard, cocoon-like puparium made from the larval skin. The adult escapes from the puparium and from the gall itself by inflating a spiny, balloon-like structure out through the front of its head. This structure presses a circular hole through both the puparium and the surface of the gall. It then retracts back into the head.

If you open a gall in the spring, you should be able to see the exit tunnel created by the larva before it pupates. Old galls usually show a small hole on the outside through which the fly escaped. Adult gall flies are about half a centimetre in length, have a light brown head and thorax, and have attractive dark patterns on otherwise clear wings.

The goldenrod gall provides the fly larva inside with both food and protection.
Drew Monkman

GETTING THROUGH THE WINTER

Depending on the species, insects can be found overwintering in every stage of the life cycle. Most insects go through complete metamorphosis (egg, larva, pupa, and adult) while a much smaller number go through incomplete metamorphosis (egg, nymph, adult). Nymphs often look like small adults, but usually don't have wings. Larvae do not look like adults and usually have a worm-like shape. Caterpillars are typical larvae. In the pupal (resting) stage, insects develop their adult characteristics such as wings but do not eat. Some species, like moths, weave a protective silk cocoon around themselves before entering the pupal stage. A butterfly pupa is called a chrysalis. There is no protective cocoon around a chrysalis.

Most insects are, of course, inactive during the winter months and enter a dormant phase called diapause. The insect's cells and tissues are protected by glycerol, a kind of natural, sweet-tasting antifreeze produced by the cells in the fall. In fact, Bernd Heinrich, in *A Year in the Maine Woods*, describes the taste of overwintering beetle larvae and carpenter ants as "glycerol-sweet."[4] Glycerol is similar to the ethylene glycol we put in our car radiators and allows an insect's body fluids to drop well below freezing without freezing solid. Not surprisingly, if you touch an insect larva at this temperature, it is still pliable. During diapause, there is no growth or development whatsoever. During "hibernation" — the term used for vertebrates — there is usually minor metabolic activity and new tissue is sometimes added to the animal's body.

The following chart provides examples of the life cycle stage at which some familiar insect species overwinter.

Insect	Life Cycle Stage	Notes
Red-legged grasshopper	Egg	In late summer, the female forces her abdomen deep into loose soil to deposit an egg pod.
Mosquitoes — genus *Aedes*	Egg	Female lays eggs at the margins of ponds. They will hatch in the spring meltwater.
Black fly	larva	Bowling pin–shaped larvae are hooked into a pad of silk, attached to wood or rocks in streams. Some black fly species overwinter as eggs.
Isabella tiger moth (woolly bear)	larva	In its well-known "woolly bear" caterpillar stage, the future moth curls up under bark, a log, etc. for the winter. It pupates in the spring, emerging two weeks later as a white Isabella moth.
Common whitetail dragonfly	nymph	Most dragonflies and damselflies spend at least a year at the bottom of pond or river in the nymph stage before transforming into aerial adults.
Mayfly	nymph	Depending on the family, some nymphs flatten themselves against solid objects in running water. Others live in *U*-shaped burrows.
Canadian tiger swallowtail	pupa	The grey pupa (chrysalis) is attached to a twig or bark by a silk button at the posterior end and a silken noose in the middle.

Tri-coloured bumblebee (*Bombus ternarius*)	adult	Pregnant queen bees overwinter in sheltered places and emerge in spring to start a new colony. Males and workers die.
Honey bee	adult	The entire colony survives the winter by clustering and vibrating their flight muscles to generate heat. They feed on stored honey.
Ladybird beetle	adult	Take refuge in the leaf litter and in crevices. Frequently found in large groups.
House mosquito (*Culex pipiens*)	adult	Mated females overwinter in sheltered locations such as animal burrows, basements and sewers. Males die before winter.
Monarch butterfly	adult	The last monarch generation of the summer migrates to the mountains of Mexico to spend the winter. Earlier generations die.

PLANTS AND FUNGI

NOVEMBER HIGHLIGHTS

All Month Long

- This is a great time of year to focus on several groups of plants of the forest floor that often escape our attention in other months. Evergreen ferns and wildflowers, club-mosses, and mosses stand out prominently against the brown leaf litter and deserve close observation. (See page 264)

- Covered by millions of fallen leaves, the forest floor is hard at work as a gigantic recycling centre. (See page 265)

- The year's flower parade has come to an end. Roadsides are now bordered by the browns and greys of dead or dormant plants. Although they appear dried and lifeless, most of these species are perennials.

- Ball-like galls are easy to see on the stems of goldenrod plants. (See November Invertebrates)

- In our woodlands, a surprising number of young trees still cling to some of their foliage. These include American beech, sugar maple, ironwood (hop-hornbeam), and oaks. Many will retain at least some leaves all winter. Trees that hold onto dead leaves are said to be "marcescent." Marcescence may benefit the trees by deterring feeding by large herbivores such as deer and moose. Dead, dry leaves are thought to make the twigs less attractive as a food source.

- During a walk in the woods, look for the rich purple leaves of bunchberry as well as the tan, wrinkled leaves and shrivelled red berries of Canada mayflower.

- The seeds and fruit of a wide variety of trees, shrubs, and vines attract birds and provide some rare November colour. The orange and yellow fruits of bittersweet and the red berries of winterberry holly and highbush cranberry are especially attractive. Ornamental varieties of crabapple, mountain ash, and hawthorn are usually heavy with fruit as well.

- Eastern hemlocks shed their seeds from late fall to late winter. It is not uncommon for the snow beneath these trees to be powdered with seeds.

- Now that the leaves have fallen from cherry trees, black knot fungus (*Apiosporina morbosa*) stands out like charred animal droppings hanging from the twigs. The dark swellings are actually the cherry's own cells that have gone amok because of the fungus.

- Late fall through early spring is a popular time to cut down trees for firewood. Before you cut, however, remember that den trees (living trees with cavities), dead trees (snags), and fallen logs are essential features of healthy wildlife habitats — not only for woodlands, but also for wetlands, hedgerows, and lakeshores. They should be left alone. Instead, try to select trees for cutting that are growing too close together, trees with obvious defects such as poor form, and undesirable species such as non-natives.

Early November

- A few non-native plants such as yarrow and dandelions may still be flowering in small numbers on lawns and roadsides.

- Oaks, tamaracks, and silver maples are the only mature, native deciduous trees that may still have some foliage. Red oaks, decked out in brownish-orange leaves, stand out in particular beside their leafless neighbours. It is therefore easy to see just how common oaks are in many areas, especially on the Canadian Shield. Some oaks will hold their leaves until the end of the month; a few will even cling to some leaves all winter long.

- Still clothed in leaves, non-native trees and shrubs remain quite conspicuous in early November. In cities, the Norway maple stands out in particular, as most turn a bright yellow and often hold their leaves up until Remembrance Day (November 11).

TURNING OUR ATTENTION TO THE NON-FLOWERING PLANTS

At first glance, a walk in the forest on a November day may seem uneventful, with little of interest to catch our attention. But now that the profusion of green leaves has retreated, one of the first things we notice is how common the various evergreen plants of the forest floor are.

Let's begin with the mosses. Take the time to get down on your hands and knees to examine these plants closely. It is like entering a miniature green forest, interspersed with strange, leafless, wiry stalks. The intricate design and distinct form, colour, and texture of the different species suddenly becomes evident. By far the most interesting characteristic of mosses lies in how they reproduce. First of all, moss really consists of two distinct generations — the familiar green, leafy plant, known technically as a gametophyte, and the wiry and leafless sporophyte with the capsule on top. When these capsules ripen, they open up and spores are dispersed. If a spore lands somewhere with sufficient moisture, it will begin to grow into a mass of green hairs that produce stems with narrow leaves — the typical "moss plant." These structures can be male, female, or sterile. Some of the plants will produce male or female sex organs among clusters of leaves at the top. Sperm produced in the male organ use a film of water in the form of rain or dew to swim to the female organ on another stem and thus fertilize the egg. The embryo then grows to form a "sporophyte," the familiar stalk with the capsule at the end. The base of the sporophyte remains anchored in the cluster of leaves. Spores develop and are released, hence repeating the cycle.

Central and eastern Ontario is home to dozens of species of mosses, many of which do not have widely accepted common names. In conifer swamps, different types of peat moss (*Sphagnum*) usually dominate. They are spongy and can form carpet-like mats. Other common groups of mosses include the upright species such as juniper hair-cap moss (*Polytrichum juniperinum*) and hair-cap moss (*Polytrichum commune*), hummock forming species like pin cushion moss (*Leucobryum glaucum*), and creeping mosses like shaggy moss (*Rhytidiadelphus triquetrus*).

Large colonies of clubmosses, too, stand out prominently in late fall, especially in the rich, shaded soils of mixed deciduous and coniferous woods. Three hundred million years ago, ancestors of these plants grew up to 30 metres high and formed a major part of the plant material that developed into coal beds. On a woodland walk in Shield country, you can expect to find ground-pine (*Lycopodium dendroideum*). With its upright, symmetrical, tree-like shape, this plant looks remarkably like a tiny pine tree. The spore-bearing leaves are tightly clustered at the tip of the stem and form a yellowish, cone-like structure. Also watch for ground-cedar (*Diphasiastrum digitatum*), whose scale-like leaves resemble those of cedar trees.

Several species of ferns are also evergreen. Probably the most common are the wood ferns (*Dryopteris* species). These lacy ferns are the greenery so often used in floral arrangements. The Christmas fern (*Polystichum acrostichoides*) is another attractive species to watch for. Its name comes from the leathery, spiny-toothed leaflets that are reminiscent of holly. It used to be a popular holiday decoration. Keep an eye open, too, for rock polypody (*Polypodium virginianum*), a small fern that grows on rocks and boulders in cool, shaded areas.

Finally, coniferous and mixed forests are also home to a variety of evergreen wildflowers. Many look as luxuriant in winter as in summer. Watch for trailing arbutus, partridgeberry, pipsissewa, twinflower, and wintergreen, to mention a few.

Ground cedar is a common clubmoss that really looks like its namesake.

Drew Monkman

NATURE'S RECYCLERS

Just a few short weeks ago our forests were ablaze in the reds, oranges, and yellows that make fall foliage so spectacular. Once the leaves are on the ground, however, we no longer give them any thought. This is a shame because, without fanfare, a profound transformation is occurring right under our feet.

The forest floor can best be thought of as a gigantic recycling centre. In this, the last stage of the ecological cycle, dead organic matter is being softened, shredded, digested, and decomposed by countless billions of organisms into simpler compounds that can be reused by the forest's plant communities. Decomposition is best thought of as a feeding process. Micro-organisms devour the organic waste from plants and animals and take from it the nutrients and energy they need to live.

Fungi are the primary decomposers of plant tissues on the forest floor and in the soil. They are capable of decomposing many of the large plant molecules that cannot be broken down by decomposers such as bacteria. Fungal spores, present in countless billions in the fall, land on the fallen leaves and, depending on conditions, may begin to germinate. However, the real action begins in the spring when the weather warms. The spores produce white or colourless thread-like strands called hyphae, which in turn secrete enzymes that render the dead leaves soft and spongy. Just 28 grams of fertile soil may contain over two kilometres of these strands. Larger organisms then take over. Millipedes and sowbugs, for example, act like grinding machines and break up a great deal of the soil litter layer.

By far the most abundant soil organisms are the numerous kinds of bacteria. A gram of soil contains billions of bacteria representing thousands of different species. Bacteria take part in almost all soil decomposition reactions. All of this biological activity in the soil forms the foundation of an extremely complicated food web that extends from miniscule decomposers like fungi and snow fleas all the way up to carnivores like owls and weasels.

The impact of earthworms in this process, however, is somewhat problematic. Worms literally eat their way through the soil by ingesting both organic and inorganic material. In this way, dead leaves are converted into rich feces known as castings. The partially digested leaf matter is thereby rendered much more susceptible to microbial breakdown and to increased nutrient release. Worms also mix the different layers of soil and aerate it through their extensive tunnel systems. There is a problem, though. Ontario's earthworms are actually a non-native, invasive species from Europe and Asia. Our forests developed in the absence of these invertebrates. Studies conducted by the University of Minnesota and forest managers have shown that these invasive earthworms are causing considerable damage.[5] Without worms, fallen leaves decompose slowly, creating a spongy layer of organic "duff." This duff layer is the natural growing environment for native woodland wildflowers. Invading earthworms eat the leaves that create the duff layer and are capable of eliminating it completely. Big trees survive, but many young seedlings perish, along with many ferns and wildflowers. It is important to never dump unwanted worms, such as those used as bait, in the woods.

WEATHER

NOVEMBER HIGHLIGHTS

All Month Long

- Another hour of daylight is lost this month. We start the month with about ten hours but finish up with little more than nine.

- The days are short and the weather is often cool and damp. However, Indian Summer conditions are also possible, with temperatures in the high teens and lots of sunshine.

- The November air is scented by damp, decomposing leaves on the forest floor and the smell of wood as the axe or chainsaw bites through a log. In more southerly parts of central and eastern Ontario, the smell of manure is often in the air as farmers spread it on their fields in the fall.

- Average precipitation in November is higher than in most other months. In Haliburton, it is actually the wettest month of the year.

- The first significant snowfall occurs and permanent snow cover is sometimes with us by month's end. Snowflakes (one or more snow crystals) form when water vapour condenses directly into ice. They are not frozen raindrops. Snowflakes are six-sided because the water molecules in a snow crystal bind together in a hexagonal shape. However, the final form the snowflake takes depends on the temperature and the humidity. At around -2°C, thin plates and stars form, while columns and slender needles appear near -5°C.

- Except for locations close to the Great Lakes, frost can be expected on about 20 days this month.

Early November

- Daylight Savings Time ends on the first Sunday of the month. Put your clocks back an hour. With the sun now setting before 5:00 p.m., darkness is often upon us by the time we come out of work. Granted, some people find the loss of light depressing. For others, however, it means cozy, relaxing evenings with no pressure to be outside working in the yard! Personally, I feel an almost spiritual connection to the cycle of the seasons and the pronounced tilt of the Northern Hemisphere away from the sun at this time of year.

November Weather Averages, 1971–2000[6]

City or Town	Daily Max. (°C)	Daily Min. (°C)	Rainfall (mm)	Snowfall (cm)	Precipitation (mm)
Owen Sound	6.5	0.5	73.9	35.5	109.4
Huntsville	4.4	-2.7	69.8	29.4	99.1
Barrie	6.1	-1.4	62.1	20.6	82.6
Haliburton	4.3	-3.1	71.8	26.4	98.1
Peterborough	5.8	-1.7	65.5	16.6	80.6
Kingston	6.8	-0.7	84.9	9.8	94.5
Ottawa	4.9	-1.9	60.5	18	77

APPROXIMATE NOVEMBER SUNRISE AND SUNSET TIMES (DST/EST)[7]

(Note: Twilight starts about 30 minutes before sunrise and continues about 30 minutes after sunset. Eastern Standard Time begins on the first Sunday in November.)

Location	Date	Sunrise (a.m.)	Sunset (p.m.)
Grey-Bruce	Nov. 1	8:00	6:14
	Nov. 15	7:19	4:57
	Nov. 30	7:38	4:46
Muskoka/Haliburton	Nov. 1	7:55	6:10
	Nov. 15	7:14	4:53
	Nov. 30	7:33	4:42
Kawartha Lakes	Nov. 1	7:49	6:04
	Nov. 15	7:08	4:47
	Nov. 30	7:27	4:36
Kingston/Ottawa	Nov. 1	7:42	5:57
	Nov. 15	7:01	4:40
	Nov. 30	7:20	4:29

NIGHT SKY

NOVEMBER HIGHLIGHTS

All Month Long

- Major constellations and stars visible (November 15 — 9:00 p.m. EST)
  Northwest: Summer Triangle made up of *Vega* (in Lyra), *Deneb* (in Cygnus), and *Altair* (in Aquila).
  Northeast: Ursa Major low in north with Ursa Minor (with *Polaris*) above; Auriga (with *Capella*) to the right and Cassiopeia near zenith; Pleiades star cluster (due east); Gemini just above horizon.
  Southeast: Orion (with *Betelgeuse* and *Rigel*) low in the east; Taurus (with *Aldebaran*) high to its right.
  Southwest: Pegasus (with the Great Square) and Andromeda (with M31 galaxy) almost at zenith.

- The Algonquian name for the full moon of November is the Beaver Moon. Beavers are indeed very active this month.

- The Pleiades (Seven Sisters) star cluster, located in the constellation Taurus, adorns the eastern sky. Viewed from the city, at least six stars are visible to the naked eye while country residents with keen eyesight may see as many as 12. Some people confuse this cluster with the Little Dipper. If anything, you might call it the "Tiny Dipper"!

Early November

- The South Taurid meteor shower peaks about November 3 and can be seen in the northeast between Taurus, Auriga, and Perseus. Under ideal conditions, it is possible to see about 15 meteors per hour.

Late November

- Around November 17, the Leonid meteor shower peaks at about 10:00 p.m. It is seen in the Leo constellation, low in the northeastern sky, and produces about 15 meteors per hour.

- Orion's arrival in the southeast seems to make November nights all the colder.

DECEMBER

The Sun Stands Still

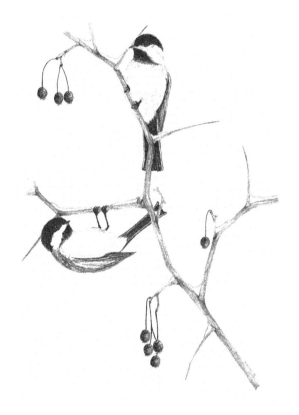

Black-capped Chickadees in hawthorn.
Kim Caldwell

So the shortest day came, and the year died,
And everywhere down the centuries of the snow-white world
Came people singing, dancing,
To drive the dark away.
They lighted candles in the winter trees;
They hung their homes with evergreen;
They burned beseeching fires all night long
To keep the year alive ...

— Susan Cooper, *The Shortest Day*

DESPITE WHAT THE CALENDAR MIGHT SAY, DECEMBER 1ST HERALDS THE BEGIN-ning of winter, a season of unadorned fundamentals. The half-hearted sun casts a pale light as it traces its lowest trajectory of the year through the southern sky. With nights as long as June days, it is no wonder that ancient peoples feared the sun's total disappearance. So, it was with great joy and relief that they celebrated the winter solstice, the day that the sun "stands still" and stops its march southward. For the next six months, the days will grow longer. In fact, the first spring birdsong is a mere seven weeks away.

With the arrival of December, naturalists look forward to the camaraderie and friendly competition of the Christmas Bird Count. Spending an entire day outside simply counting and looking for birds is the ideal antidote to the pressures, excesses, and mad rush of the holiday season.

As attractive as our holiday decorations may be, nature itself provides the most enduring Yuletide adornments: festive winterberry holly fruits, radiant cardinals, fluffy chickadee baubles, hoar-frosted windows, and shimmering icicles. The Christmas tree, symbolic of life's vigour in the face of winter, fills our homes with the resinous fragrance of the northern forest. Nature also supplies its own array of sounds for the festive season — the cracking and rumbling of new ice on the lakes, the shrill scolding of red squirrels, and the croaking of ravens patrolling the Shield. Although December sees the year come full circle, it is neither an end nor a beginning. Like every other month, it is simply part of an indivisible whole.

BIRDS

DECEMBER HIGHLIGHTS

All Month Long

- Almost all migratory birds that breed in central and eastern Ontario are now on their wintering grounds, where they must compete with the resident bird community of the area for food resources. (See page 276)

- The number of species and individual birds visiting feeders in winter often fluctuates dramatically. People often ask, "Where are all the birds?" (See page 273)

- As a general rule, winter birds are most often found around human habitations, open water, thick stands of coniferous trees, shrubby areas, and farmland. In Shield country, keep an eye open for finches foraging right on the road, attracted by the grit used in sanding operations. In the city, watch the sky for Cooper's and Sharp-shinned Hawks. They are fairly common in the winter and can be identified at considerable distances by their unique "flap-flap-glide" flight pattern.

- December is the peak calling period for both Eastern Screech and Great Horned Owls. The best nights to hear them are usually those with falling barometric pressure and a full or gibbous moon.

- In late fall and early winter, rare avian visitors to central and eastern Ontario may unexpectedly show up at feeders. Some, such as the Varied Thrush, a species native to western North America and the Pacific coast, are thousands of kilometres from their normal range.

Early December

- Ducks lingering until freeze-up may include Common Goldeneye, Common and Hooded Mergansers, and American Black Duck. A small number of Common Loons, mostly young-of-the-year birds, remain until the ice comes, as well. In some areas where there is open water, large resident Mallard populations will also remain all winter long.

- Loons sometimes become trapped in the ice when a pool of open water freezes overnight. Even if some open water remains, loons require a large expanse on which to run in order to take off.

- Birders often head to Moses-Saunders Power Dam in Cornwall and to Niagara Falls in early December. The entire Niagara River corridor is internationally recognized as a "Globally Significant Important Bird Area," as it has the largest and most diverse concentration of gulls in the world.

Mid-December

- From December 14 to January 5 (inclusive), Christmas Bird Counts take place throughout North and South America. The count data reflects trends in bird populations such as the expansion of Red-bellied Woodpeckers into many areas of central and eastern Ontario. (See page 275)

WHY IS THERE NOTHING AT MY FEEDER?

Many people often wonder why the number of birds visiting their feeder fluctuates so much from one month or week to the next. There are several possible explanations. Winter finches such as siskins and redpolls, for example, are highly nomadic and will wander great distances over the course of the winter. Human offerings of seed are never enough to make them stay, since the availability of natural food is the determining factor. Even at the best of times, birds spend no more than 25 percent of their feeding day at feeders, preferring to seek out whatever wild food is available the rest of the time. This percentage is even lower during years when wild food is particularly abundant, such as in the winter of 2011–12. Second, many species such as House Finches, American Goldfinches, and sometimes even Northern Cardinals now travel in flocks and may frequent only a small number of feeders. If your feeder is not on their list, you may find things very quiet. Also, with the number of people feeding birds these days, there are relatively fewer birds to go around. Finally, the presence of a raptor in the neighbourhood may also explain why, on a given day, there are fewer birds. Hawks, of course, are drawn by the fact that people are feeding birds in the first place.

I always feel privileged when a Sharp-shinned or Cooper's Hawk makes an appearance in my yard. In addition to being beautiful birds in their own right, they allow me to observe first-hand the food chain at work. I remember one Christmas morning when a Cooper's Hawk killed a Mourning Dove before our very eyes as we unwrapped presents. I explained to my children — who were understandably rather upset — how lucky we were to be able to actually witness the entire food chain in action as the sun's energy passed from sunflower seed to dove to hawk.

Immature Cooper's Hawks are brown and white with thin, dark streaks on the breast.

Karl Egressy

COUNTING BIRDS AT CHRISTMAS

For birders, December and early January is a favourite time of year. Not only is it the holiday season, but Christmas Bird Counts (CBCs) are taking place all across Ontario. A count is held during a single calendar day and covers the area of a circle measuring 24 kilometres in diameter. The circle is divided up into a number of sectors, with each sector being covered by a different group of observers. The birders are out from dawn until dusk recording not only the different species seen but also the number of individual birds. Some observers actually start before dawn in the hope of hearing owls. Considerable time is also spent tactfully peering into people's yards to see what is coming to their feeder. You do not have to be an expert to take part in a bird count. Simply having extra eyes and ears in the group invariably leads to more birds being counted. You can also participate as a "feeder watcher." If you can devote at least two hours to watching your feeder on count day, your sightings can be phoned in and included.

Depending on what sort of weather has prevailed during the fall and on the availability of natural foods such as berries and cones, the number of species counted changes considerably from year to year. A mild fall means more species, since some birds will linger in our area before heading south or perishing. A heavy wild grape crop will probably allow greater numbers of robins and waxwings to overwinter, while plentiful pine and spruce cones might attract various species of winter finches. The weather on the day of the count can have a major impact, as well, especially if there is heavy snow or rain. On average, the Peterborough count records about 50 species, but counts on Lake Ontario, for example those in Kingston, often tally more than 100 different species. In comparison, a CBC held in Panama each year usually comes up with over 300 species! The count results are submitted to the National Audubon Society in Washington, D.C. The data, collected over a long period of time, can provide valuable information on the relative abundance and distribution of bird species. A Christmas Bird Count, however, is other things as well. It serves as a welcome break from the stress and consumerism of the holiday season and as an excuse to get together with other birders for some friendly competition and camaraderie.

Some Commonly Recorded Species on Central and Eastern Ontario Christmas Bird Counts

(List organized by family)

Mallard*	Red-breasted Nuthatch
Common Goldeneye*	White-breasted Nuthatch
Common Merganser*	Brown Creeper
Sharp-shinned Hawk*	Golden-crowned Kinglet
Cooper's Hawk*	American Robin*

Red-tailed Hawk*
Bald Eagle
Ruffed Grouse
Ring-billed Gull*
Herring Gull*
Rock Pigeon
Mourning Dove*
Great Horned Owl
Barred Owl
Downy Woodpecker
Hairy Woodpecker
Pileated Woodpecker
Northern Shrike
Blue Jay
American Crow
Common Raven
Black-capped Chickadee

European Starling
Cedar Waxwing
American Tree Sparrow
Dark-eyed Junco
Snow Bunting
Northern Cardinal*
Pine Grosbeak**
White-winged Crossbill **
Red Crossbill **
House Finch*
Purple Finch**
Pine Siskin**
Common Redpoll**
Hoary Redpoll**
American Goldfinch
Evening Grosbeak**
House Sparrow

usually restricted to counts south of the Canadian Shield
*** numbers fluctuate greatly from year-to-year. Can be completely absent.*

MAKING A LIVING IN SOUTHERN CLIMES

When it comes to coping with winter, most bird species that breed in central and eastern Ontario simply pack up and leave. Some go no farther than the southern United States, while others — the so-called neotropical migrants — head to the West Indies, southern Mexico, Central America, and even South America. Many of our backyard American Robins this month are hanging out in South Carolina, while the Barn Swallows that nested in the barn are scattered from Costa Rica to Uruguay. The hummingbirds that came to the feeder all summer are now anywhere from southern Mexico to Panama, after having made a non-stop 18- to 24-hour crossing of the Gulf of Mexico. One of our best-loved species, the Common Loon, winters mostly on salt water in an area extending from Newfoundland to Mexico.

We might assume that the birds we know so well from summer pursue a similar lifestyle in the winter. This is often not the case, however. In fact, many species can seem like completely different creatures on their wintering grounds. Whereas in the summer, an American Redstart might be typically found feeding along a bucolic forest edge in cottage country, the winter version of the same bird can just as easily be seen catching flies in a Cuban barnyard or around an outdoor privy.

Practically all habitat types available in the neotropics are used by North American migrants in the winter. These can range from rainforest and mangrove to suburban backyards and even shade-grown coffee plantations. In this traditional way of growing coffee, the coffee bushes are planted in and around the forest canopy where they are protected from the sun. By buying shade-grown coffee from Latin America, you are helping to save crucial habitat for everything from Rose-breasted Grosbeaks and Swainson's Thrushes to Latin America resident species like Blue-crowned Motmots and Lineated Woodpeckers. Other species, like Black-and-white Warblers, are generalists and occupy many kinds of wintering habitat.

Most neotropical migrants, however, have fairly tight restrictions on what constitutes a suitable environment. This puts them at risk from habitat destruction such as deforestation. For example, the winter range of the Cerulean Warbler, a rare species in most of central and eastern Ontario, is limited to the eastern Andean foothills of Colombia and neighbouring countries. Conversion to agriculture is putting the cerulean's future in jeopardy.

Still, there is some good news. Many migrants eat a diet high in fruit and nectar during the winter. Therefore, large concentrations of our songbirds are often found in areas where fruit trees prosper. Several studies have found that disturbed habitats with young forests offer the most fruit availability. Baltimore Orioles and Scarlet Tanagers, for example, are attracted to fig and Cecropia trees in these habitats.

As a last word, we are probably being presumptuous to think of neotropical species as "our" birds. It's much more accurate to think of hummingbirds, orioles, and warblers as tropical species that simply make a brief trip northward in the summer to breed before heading home again for the other eight months of the year.

WHERE ARE THEY NOW?

When the National Audubon Society publishes the results of the Christmas Bird Counts each year, it is interesting to note some of the locations where large concentrations of "our" birds spend the winter months.[1]

(List organized by family)

Common Loon: 832, Wilmington, North Carolina
Common Merganser: 39,640, Point Pelee, Ontario
Osprey: 350, Cocoa, Florida
Eastern Phoebe: 1,040, Matagorda County — Mad Island Marsh, Texas
American Crow: 207,000, Middle Fork River Valley, Illinois
Eastern Bluebird: 770, Ragersville, Ohio
Swainson's Thrush: 95, Mindo-Tandayapa, Ecuador
Chestnut-sided Warbler: 321, La Selva, Lower Braulio Carrillo National Park, Costa Rica

White-throated Sparrow: 3,989, Lower Kent County, Maryland
Red-winged Blackbird: 2,800,000, Sooner Lake, Oklahoma
Baltimore Oriole: 136, La Selva, Costa Rica

Mammals

December Highlights

All Month Long

- "Nip twigs" on the ground below conifers are another sure sign of red squirrel activity. If you walk quietly through the woods, you will sometimes even hear the sounds of the squirrels tearing cones apart with their teeth. (See page 279)

- Bears are snuggly ensconced in their winter dens. (See page 279)

- The red squirrel's coat is now a brilliant russet. It is also much thicker than the summer coat and will provide better protection from the cold.

- Throughout the late fall and winter, gray squirrels are often seen high up in Manitoba and Norway maples feeding on the keys. Nearly all gray squirrels in southern and central Ontario are the melanistic (black) form.

- Like most mammals, the fur of the fisher varies seasonally, becoming denser and glossier in the winter months. Fisher pelts were in such demand that they almost became extinct in the early 1900s. Since then, fisher populations have increased naturally due to a number of factors, including more forest habitat and lower snow depths. The species has now expanded it range well south of the Canadian Shield with records in Durham Region and in the vicinity of — if not actually in — Presqu'ile Provincial Park.[2]

- Identifying and interpreting mammal tracks is a fascinating pastime and adds a great deal to a winter outing. Also keep an eye open for mammal scat (animal droppings), which is actually easiest to identify in winter. Coyote scat, for example, is often found on rail-trails or roads. It contains mostly hair and is about the diameter of a cigar. It is usually black and tapered at one end. However, if it has been lying in the sun for a while it will be bleached to a lighter colour.

- If you are fortunate, you may have a family of flying squirrels providing nightly entertainment at your bird feeder. They are usually quite tame.

Flying squirrels have dark, bulging eyes that provide excellent night vision.
Ken Thomas, Wikimedia

Taking a Nip to Get Through the Winter

In late fall and winter, red squirrels are a common sight in mixed and coniferous woodlands. In addition to their agitated calls, red squirrels leave behind a number of other signs of their presence. Among these are "nip twigs" scattered on the ground under spruce and other conifers. In order to get at the cones and terminal buds that are an integral part of their diet, squirrels nip off the tips of conifer branches, allowing the twig to fall to the ground. They then scurry down from the tree and remove the cones and buds, leaving the rest of the twig on the ground. While the buds are consumed immediately, the cones are often stored in hollow logs, under the roots of dead trees, and in spaces between rocks. They are later taken to a favourite eating spot such as a tree stump. Only the seeds, however, are actually consumed. A pile of cone scales and shafts accumulates at the eating site. This pile is referred to as a "midden," and can sometimes measure half a metre deep!

Extraordinary Bear Hibernation

In winter, a sleeping bear's heart rate drops to about eight beats a minute from a normal sleeping rate of 40. Breathing, too, slows to only several breaths a minute. However, the animal's body temperature only cools down by a few degrees. A true hibernator such as a groundhog experiences body temperatures close to freezing and actually enters a frigid, deathlike state from which it cannot be easily awoken. A sleeping bear, on the other hand, can be aroused even by small noises. The true miracle of bear hibernation, however, lies in the animal's unique chemistry. First of all, bears do not urinate during the winter. For most animals, this would result in fatal poisoning from the buildup of urea. A sleeping bruin, however, burns fat rather than protein. It therefore produces very little urea in the first place. The urea that is produced is broken down, and the nitrogen is used to build protein. The ability to build protein while fasting allows black bears to maintain their muscle and organ tissue throughout the winter. Without urination or defecation, the only water loss is through the animal's shallow breathing. Even this water, though, is quickly replaced from the bear's fat stores. Another amazing adaptation is that hibernating bears do not appear to suffer from any signs of osteoporosis (loss of bone mass), even after months of inactivity. Clearly, all of these special adaptations on the part of bears have research implications when it comes to improving human health and fitness. Some say that bears would make the ideal space travellers: eat from July through October and live off fat for the rest of the year!

AMPHIBIANS AND REPTILES

DECEMBER HIGHLIGHTS

All Month Long

- Mudpuppies — foot-long, permanently-aquatic salamanders — are active and feed all winter long. They retain the gills and smooth skin of larvae as adults, but go undetected in many water bodies because of their secretive habits. Kemptville Creek at Oxford Mills is probably the best place to see mudpuppies in all of central and eastern Ontario. They can be observed here in large numbers throughout the winter months. Nowhere else can you see active amphibians when the air temperature is -26°C!

FISH

DECEMBER HIGHLIGHTS

All Month Long

- Lake Sturgeon, a highly migratory fish species that is dependent on river environments, move into deeper water for the winter months. Some sturgeon find suitable deep water in pools of rivers such as the Nottawasaga, while others move out into the Great Lakes. Fish use less energy in these areas of slower water, which helps them survive the winter. However, being migratory makes sturgeon vulnerable to human influences such as the construction of artificial barriers or manipulation of water flow. Like a number of other fishes, they are now designated as a Species at Risk in Ontario. (See below)

The American Eel lives mainly along the St. Lawrence River, Lake Ontario, and their tributaries. Historically it was present throughout the Ottawa River drainage system.
U.S. Fish and Wildlife Service, Wikimedia

FISH IN DECLINE

At least 11 species of fish that are native to central and eastern Ontario are now classified as Species at Risk. This includes the American Eel, Bridle Shiner, Channel Darter, Cutlip Minnow, Grass Pickerel, Lake Sturgeon, Northern Brook Lamprey, Pugnose Shiner, Redside Dace, River Redhorse, and Shortnose Cisco. The introduction of species that are not native to a given ecosystem is one of the major threats to fish biodiversity. This can happen when fish are moved either accidentally (e.g., movement through canals) or deliberately (e.g., releasing unused bait fish) from one body of water to another. A fish species that is new to a given body of water can compete with native fish species for resources or even become a predator upon them. Moving native species from one water body to another can also be problematic (and is illegal). Moving fish can also result in the spread of pathogens such as viral hemorrhagic septicemia (VHS). It has been responsible for a number of large die-offs in the Great Lakes and the St. Lawrence River in recent years.

The construction of dams, too, is a major threat to fish populations because it can interrupt seasonal spawning movements. Dams have had an especially negative impact on Lake Sturgeon populations. Other common causes of fish decline include overfishing and the degradation of aquatic habitat by activities such as land-use practices and shore-line alterations. The invasion of fish habitat by non-native plants, such as milfoil, is also a problem. Milfoil crowds out the native plants that some fish prefer.

One of the most tragic fish decline stories is that of the American Eel. Eel populations have declined precipitously both in Ontario and throughout their global range. For example, the number of fish that use the eel ladder at Cornwall has decreased by more than 90 percent since the 1980s.[3] In addition to factors such as over-fishing and the impact of dams on eel migrations, changes to the ocean currents that aid the distribution of larval eel may also have had an influence on their abundance in the northern portion of their range.[4]

INSECTS AND OTHER INVERTEBRATES

DECEMBER HIGHLIGHTS

All Month Long

- Pregnant adult queen wasps overwinter in crevices in rocks and wood. You may inadvertently bring one into the house, tucked away in a piece of Yuletide firewood. Be sure to buy and burn all firewood locally, because of the danger of inadvertently spreading invasive insect species such as the emerald ash borer and the Asian long-horned beetle. (See below)

INVASIVE INSECTS[5]

Late fall is often a time that people purchase firewood. However, because of the danger of accidentally introducing invasive insects into new areas of the province, firewood should always be obtained locally, burned onsite, and not moved to new areas. The emerald ash borer (*Agrilus planipennis*) has now spread from southwestern Ontario to Toronto and Ottawa. Encroaching from the south, east, and west, it might only be a matter of time before the insect finds its way to cottage country where it could potentially devastate the ash tree population. Adult borer beetles are metallic green, slender, and only measure from about nine to 14 millimetres in length. In the caterpillar-like larval form, the borer feeds in an *S*-shaped pattern just under the bark of ash trees. This disrupts the tree's transportation of water and nutrients and will effectively kill the tree within two or three years. Top branches usually die off first. After overwintering in the tree, the larvae transform into adult beetles and chew their way out of the tree through *D*-shaped exit holes. All native ash species and sizes of tree are susceptible to invasion.

As if emerald ash borers were not enough, Ontario is also facing an assault from another invasive species — the Asian long-horned beetle (*Anoplophora glabripennis*). This insect attacks and kills a wide variety of hardwood trees, including all species of maple. Conifers are spared, however. A large, robust insect, the adult measures 20 to 35 millimetres in length and seven to 12 millimetres wide. It is shiny black, with up to 20 white dots on its back. Be careful not to confuse it with the somewhat similar but native white-spotted sawyer (*Monochamus scutellatus scutellatus*), which is a common borer in dead or dying conifers. The latter has diffuse, indistinct white markings. As with ash borers, larval Asian long-horn feeding activity under the bark eventually cuts off the transport of nutrients and water to the tree. Infested trees will demonstrate premature leaf drop and generalized crown dieback. Adults feed on the leaves of the trees. Asian long-horned beetles can overwinter in infected trees as eggs, larvae, or pupae. This species showed up in Toronto in 2003; however, thanks to aggressive control measures, it has not yet become widespread in North America. To report signs and symptoms of trees infested by invasive insects, contact the OMNR at 1-800-667-1940.

PLANTS AND FUNGI

DECEMBER HIGHLIGHTS

All Month Long

- Like animals, wildflowers use different strategies to survive winter. Annuals, biennials, and perennials have all developed their own unique adaptations. (See below)

- Balsam fir makes the perfect Christmas tree — symmetrical shape, long-lasting needles, and a wonderful fragrance. (See page 284)

- One of the oldest holiday traditions is to decorate the home with English holly (*Ilex aquifolium*), a European species with evergreen, spine-edged leaves. Central and eastern Ontario actually has its own native holly, namely the deciduous winterberry holly (*Ilex verticillata*). Its bright red berries are one of the visual treats of wetland edges in the late fall and are now very popular themselves as Christmas decorations.

- Holiday wreaths and other floral decorations bring many aspects of nature into our homes at this time of year. These often include fragrant pine, fir, spruce, and cedar boughs, fruit-laden winterberry holly twigs, bright red dogwood branches, and a variety of different cones.

- With all the leaves off our deciduous trees and evergreen trees and boughs so much a part of the holiday season, this is a great month to learn to identify our native conifers. Mnemonics such as "white pine needles come in bundles of five and *white* has five letters" can be helpful.

- Winter is a good time to look at the many different bracket fungi (polypores) that grow out flat from tree trunks. (See September Plants and Fungi)

PLANT STRATEGIES FOR WINTER SURVIVAL

Like animals, plants have developed special strategies for winter survival. Wildflowers in particular display some of the most diverse adaptations. Like garden plants, wildflowers can be annuals, biennials, or perennials. Annuals complete their life cycle in a single growing season and prefer open, full-sun environments. They are especially abundant in sites that have been recently disturbed. Some common native annuals include common ragweed and beggar's tick. The majority of annuals, however, are non-native and include lamb's quarters, crab grass, and various mustards. With the first heavy frosts, most annuals die, leaving only their seeds to survive the winter. The seeds will not germinate until they have been frozen for an extended period of time. This prevents fall germination and certain death from the cold. The seeds are long-lived and can remain dormant in the soil for years. The annuals of one group, known as winter annuals, set seed in the fall and survive the winter as a prostrate rosette of ground-hugging leaves that flower in the spring, produce seed, and start the cycle again. The daisy fleabane is an example of a native winter annual.

Biennials take two years to produce their seeds. The first year is spent putting down roots and forming a rosette. This ring of ground-hugging leaves survives under the snow and sends up a stem early in the spring, allowing for a quick start on the growing season. Biennials inhabit open full-sun environments such as old fields and include the native black-eyed Susan and evening primrose, as well as common aliens such as Queen Anne's lace and common burdock. Relatively few species of plants are biennials.

The vast majority of our wildflowers are perennials. Although the aboveground parts of many perennials die in the fall, the plant continues to survive below the ground, usually as a root, rhizome, or tuber. Perennials often have a prominent tap root or a dense, fibrous root system. A rhizome is a root-like underground stem that produces new roots and stems, allowing the plant to spread. A tuber is a short, thick underground stem. These various root-like structures continue to produce new plants for many years. After early domination by annuals and biennials, perennials soon take over field habitats. Asters, goldenrods, and milkweeds are three of the most common native perennials in fields. In our woodlands, two of the best-known groups are the trilliums and orchids. (See May Plants and Fungi)

The Best Christmas Tree

Among the many tasks to be completed during the holiday season is the purchase of a Christmas tree. What species do you choose? For many, the best choice is the traditional favourite, the balsam fir. The long-lasting, dark-green needles exude a wonderful balsamic fragrance that immediately evokes memories of Christmases past. The fir is also one of the most perfectly symmetrical evergreens and probably the easiest species to decorate. Although you may have to pay a little more for this species, it is worth the extra expense.

There are many advantages to buying a real tree, too. Christmas tree farms often provide good wildlife habitat for a number of bird species and are often grown on soils that could not support other crops. Buying your tree from a local farm supports the local economy. In most municipalities, they can also be recycled into mulch and compost. Personally, I like to put the tree outside beneath a bird feeder to provide additional cover for ground-feeding species. As for artificial trees, they are reusable, but don't forget that they are usually made with polyvinyl chloride (PVC), which is one of the most environmentally offensive forms of petroleum-derived plastic. In addition, they are almost always made overseas and shipped great distances, thereby creating a huge carbon footprint. If you really want to get creative and have conifers growing in your yard, why not consider going outside and decorating one of them?

The balsam fir has flat, dark-green needles with a rounded or notched tip and a whitish underside.

Drew Monkman

WEATHER

DECEMBER HIGHLIGHTS

All Month Long

- Welcome to the "dark turn of the year." Daylight this month averages only about 8 ¾ hours. Compare this to 15 ½ hours in June — a difference of 6 ¾ hours!

- The western third of central Ontario is located in the snowbelt, a term describing localities east of Lake Huron and Georgian Bay where heavy snowfall in the form of lake-effect snow occurs. (See page 287)

- The different winter air masses that affect central and eastern Ontario cause considerable variability in winter weather. Air masses originating over the Pacific tend to be cool and raw, while Arctic air brings frigidly cold conditions.

- Central and eastern Ontario is subject to episodes of freezing rain most winters. However, it is usually confined to a small area. Freezing rain is like ordinary rain until it strikes a frozen surface, where it forms a layer of clear ice. As the layer thickens, the weight of the ice can cause great damage to forests as branches break off and entire trees are brought down. We frequently see the devastation from freezing rain events, often without recognizing it, when we drive or walk through wooded habitats such as the area east of Kaladar on Highway 7. The "Great Ice Storm" of 1998 was a freezing rain storm.

- Conifer needles exude wonderful resinous smells, especially on sunny, mild December days. Indoors, a balsam fir Christmas tree scents the air like no other.

- Before too much snow falls, this is good time to walk around the edge of swamps to look for interesting ice formations such as ice crystals imitating stalagmites in a limestone cave. Leaves, sticks, and bubbles frozen in the ice can also be intriguing.

Mid-December
- During most years, lakes are frozen by mid-December. However, there is considerable variability in the date of freeze-up from one year to the next. (See page 287)

Late December
- The winter solstice marks the shortest day of the year as the sun traces its lowest and shortest arc through the sky. It is also the official beginning of winter. (See page 288)

- Even though the days do grow longer after the solstice, they begin to do so very slowly. The increase in daylight is in the afternoon. On December 31, sunset is only about six minutes later than it was on December 21, but, more surprisingly, the sun also rises about three minutes *later* at month's end than it did on the solstice.

- The chance of having a white Christmas in most of central and eastern Ontario is over 80 percent.

DECEMBER WEATHER AVERAGES, 1971–2000[6]

City or Town	Daily Max. (°C)	Daily Min. (°C)	Rainfall (mm)	Snowfall (cm)	Precipitation (mm)
Owen Sound	0.7	-5.4	32.9	93.9	126.9
Huntsville	-1.9	-10.2	21.2	71.9	93.1
Barrie	0	-7.9	21.3	62.4	83.7
Haliburton	-2.5	-11.6	28	64.9	92.9
Peterborough	-0.7	-8.8	34.3	42.4	73.8
Kingston	0.4	-8.1	54.3	47.6	99
Ottawa	-2.9	-10.3	28.8	52.2	74.1

THE LAKES FREEZE OVER

Central and eastern Ontario lakes are usually frozen by mid-December. The speed at which freezing can occur is amazing. One day the whole lake is open; the next it is completely frozen. There is no intermediate stage between water and ice — it is one or the other. How this comes about is interesting. When water cools to 4°C, something remarkable occurs. Instead of becoming denser and sinking, it actually begins to expand and becomes lighter. Being colder — and lighter — than the water below, it stays on the surface of the lake. As the water is cooled even more, the expansion continues until 0°C is reached. At this point, the molecules lock into the pattern of a solid and form ice. In his book *Winter*, Doug Sadler writes: "To the physicist as to the layperson, the suddenness of the metamorphosis is truly astonishing. It is as if tension builds up to an unbearable point where it has to be released in a sudden, orgasmic moment of creation."[7]

This process of molecular bonding and ice formation goes ahead even in the presence of waves. The expansion of the water from 4°C up to the point of freezing means that the ice is 10 percent lighter than the water below and therefore floats on the surface. The first ice that develops is called "black ice." It is highly transparent. The weight of the snow that accumulates on the black ice causes it to crack and to allow flooding of the ice surface and snow cover. When this "slush" refreezes, opaque "white ice" is formed. More snow then accumulates on the white ice, which together insulate the lake from the colder atmospheric temperatures above and slow the development of more black ice.[8]

LAKE-EFFECT SNOW

As people living east and south of Lake Huron and Georgian Bay can attest, colliding pressure systems are not always necessary to produce snowstorms. This region also receives frequent heavy lake-effect snow squalls that contribute 30 to 50 percent of the annual winter snowfall and increase seasonal snowfall totals to upwards of three metres!

During the summer season, the Great Lakes absorb large amounts of heat. Because water heats slowly but retains its stored heat for a long time, the open waters of the Great Lakes are much warmer than the Arctic air that crosses them, especially during the late fall and early winter. This makes November and December the prime months for heavy lake-effect snows.

When cold Arctic air inevitably sweeps over the relatively warmer lake water, the water heats the air mass's bottom layer. At the same time, water from the surface of the lake evaporates into the cold air. Since warm air is lighter or less dense than cold air, the heated air rises and begins to cool. As the air cools, the moisture that evaporated into it condenses and forms clouds and snow begins falling on the leeward shores. Precipitation is heaviest in areas where the land quickly rises in elevation: about 17 centimetres for every 30-metre rise.[9] Large amounts of snow can fall in a very short time.

The areas of central and eastern Ontario affected by storms of this kind are known as snowbelts, and include the Bruce Peninsula, Owen Sound, Parry Sound, Huntsville, and Barrie. January snowfall is more than twice what areas just east of the snowbelt — the Kawarthas, for example — receive. Total winter snowfall in snowbelt areas can reach 300 to 400 centimetres.

THE MYSTIQUE OF THE WINTER SOLSTICE

Caught up in the mad rush of the holiday season, most of us are unaware that a profound celestial event takes place this month. December is the month of the winter solstice — the shortest day of the year and the first official day of winter. At the solstice, the northern hemisphere is tipped farthest away from the sun. From our perspective, we see the sun tracing its lowest arc through the sky. Even at noon, it remains relatively low in the southern sky. Because the sun rises at its southernmost point on the eastern horizon and sets at its southernmost point in the west, its trajectory through the sky is also much shorter. This, in turn, results in an extremely short day. At the solstice, Peterborough receives only eight hours and 51 minutes of daylight. Compare this to the 15 hours and 32 minutes of daylight at the summer solstice on June 21!

The solstice has always been a time of awe and amazement. It is an event that was noticed and celebrated by cultures all over the world and, in the opinion of some, was a precursor to faith. As ancient peoples would watch the sun rise and set farther and farther south each day and notice the hours of daylight grow shorter, they would almost certainly fear the sun's complete disappearance. But, just when the world appeared to be on the brink of utter darkness and oblivion, the sun would suddenly stop its southward march in sunrise and sunset points; its noontime elevation, too, would cease to descend lower and lower in the sky. It would essentially "stand still" for several days, before once again proceeding to move northward and to climb higher and higher in the sky. The joy and reverence that the ancients would have felt as this unfolded are not hard to understand. The celebration of the solstice existed in many cultures. For people of northern climes, the winter solstice represents the assurance that the days are once again growing longer and that spring will indeed return. The solstice also reminds us of the close links between the celebrations of the holiday season and the rhythms of the natural world.

APPROXIMATE DECEMBER SUNRISE AND SUNSET TIMES (DST)[10]

(Note: Twilight starts about 30 minutes before sunrise and continues about 30 minutes after sunset.)

Location	Date	Sunrise (a.m.)	Sunset (p.m.)
Grey-Bruce	Dec. 1	7:39	4:46
	Dec. 15	7:53	4:44
	Dec. 21 (solstice)	7:57	4:46
	Dec. 31	8:00	4:52

Muskoka/Haliburton	Dec. 1	7:34	4:41
	Dec. 15	7:47	4:40
	Dec. 21 (solstice)	7:51	4:42
	Dec. 31	7:55	4:48
Kawartha Lakes	Dec. 1	7:28	4:36
	Dec. 15	7:42	4:34
	Dec. 21 (solstice)	7:46	4:36
	Dec. 31	7:49	4:43
Kingston/Ottawa	Dec. 1	7:21	4:29
	Dec. 15	7:34	4:27
	Dec. 21 (solstice)	7:38	4:29
	Dec. 31	7:42	4:36

NIGHT SKY

DECEMBER HIGHLIGHTS

All Month Long

- Major constellations and stars visible (December 15 — 9:00 p.m. EST)
  Northwest: Cassiopeia near zenith; Summer Triangle with *Vega* and *Altair* just above horizon and *Deneb* in mid-sky.
  Northeast: Ursa Major low in the north with Ursa Minor (with *Polaris*) above it; Auriga (with *Capella*) near zenith; Gemini (with *Castor* and *Pollux*) in mid-sky.
  Southeast: Pleiades near zenith; Orion (with *Betelgeuse* and *Rigel*), Taurus (with *Aldebaran*) high to its right; Canis Major (with *Sirius*) low to its left; Auriga (with *Capella*) below Taurus.
  Southwest: Pegasus (with the Great Square) high in the west; Andromeda (with M31 galaxy) almost at zenith.

- The Algonquian name for the full moon of December is the Cold Moon.

- The early morning hours of December 13 and 14 are the peak viewing days for the Geminids meteor shower. Considered to be the most consistently good meteor shower of the year, the Geminids are known for producing up to 60 meteors per hour at their peak. Some meteors should also be visible as early as December 6 and as late as December 17.

- Seeing the northern lights, or aurora borealis, is a rare but memorable experience. Although they can occur at any time of the year, somehow the northern lights always seem more impressive on a cold winter night.

- The winter constellations shine brightly and are easy to pick out. In the southeast, look for the Winter Six: Taurus, Gemini, Auriga, Canis Major, Canis Minor, and, of course, Orion. The most prominent stars in these same constellations form an asterism known as the Winter Hexagon. You can see it by imagining a line joining Rigel, Aldebaran, Capella, Pollux/Castor, Procyon, and Sirius.

- The December moon rises about 30 degrees north of due east and sets 30 degrees north of due west. It also rides higher in the sky than during any other month of the year.

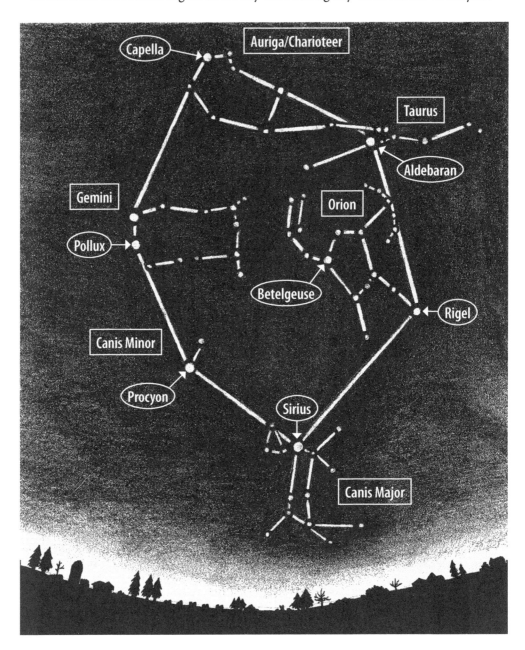

The Winter Six constellations and the asterism known as the Winter Hexagon.
Jean-Paul Efford

ORION, THE RULER OF THE WINTER SKY

Few constellations are as conspicuous and easy to remember as Orion, the Hunter. Maybe this is because Orion is one constellation that actually looks like its namesake. The belt, sword, shield and club are all easy to see and imagine. Of the four bright stars that form Orion's shoulders and knees, two are particularly impressive. Betelgeuse, which forms the left shoulder, is a reddish supergiant, 600 times the diameter of the sun. The lower right corner of Orion's torso is occupied by Rigel, a magnificent bluish-white star 50,000 times as luminous as our sun. However, the most fascinating feature of this huge constellation is the Orion Nebula, located in the "sword" hanging from Orion's belt. It is an area of the galaxy where new stars are forming and is spectacular in even a small spotting scope or telescope. Like the Big Dipper, Orion is useful in locating other stars and constellations. The belt acts as a pointer in two directions. To the right, it points to Aldebaran in Taurus, the Bull, and then on toward the Pleiades. To the left, the belt points to Sirius in Canis Major; Sirius is the brightest of all the stars. A line extended from Rigel through Betelgeuse points to Gemini, the Twins (Pollux and Castor).

APPENDIX 1

At-a-Glance Seasonal Abundance for Some Commonly Seen Birds

The following line charts attempt to show as accurately and clearly as possible the seasonal abundance patterns of some of the bird species most often seen in central and eastern Ontario. Each bird has been given a status based on my judgement of how easily it would be found by someone searching in appropriate habitat. The charts are most accurate for inland locations along the edge of the Canadian Shield but do apply to the region as a whole. Waterfowl and shorebird (e.g., sandpipers) abundance varies considerably, depending on the migration route the birds follow and the availability of appropriate habitat. Snow Geese, for example, tend to migrate along the Ottawa River corridor where they are often abundant. They may, however, be completely absent elsewhere. Note, too, that not all species are seen every year. Please refer to the Atlas of the Breeding Birds of Ontario: 2001 – 2005 for detailed information on the breeding range of a given species.

Common - likely to be seen every day; may be present in large numbers *(e.g., American Robin)* OR in lesser numbers but throughout the area *(e.g., Common Loon)*

Uncommon - not expected daily, but seen with repeated visits *(e.g., Pileated Woodpecker)*. May be secretive *(e.g., Brown Creeper)*

Rare - if present, always in low numbers and usually hard to find

Irregular - present irregularly, varying from absent to common *(e.g., Common Redpoll)*

(Ottawa area) Restricted range - generally restricted to specified region, although occasionally turns up elsewhere

★ Breeding - nests in central and/or eastern Ontario *(e.g., Bald Eagle)*

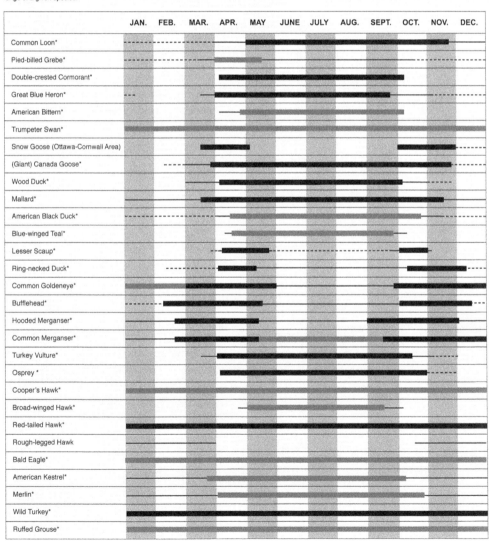

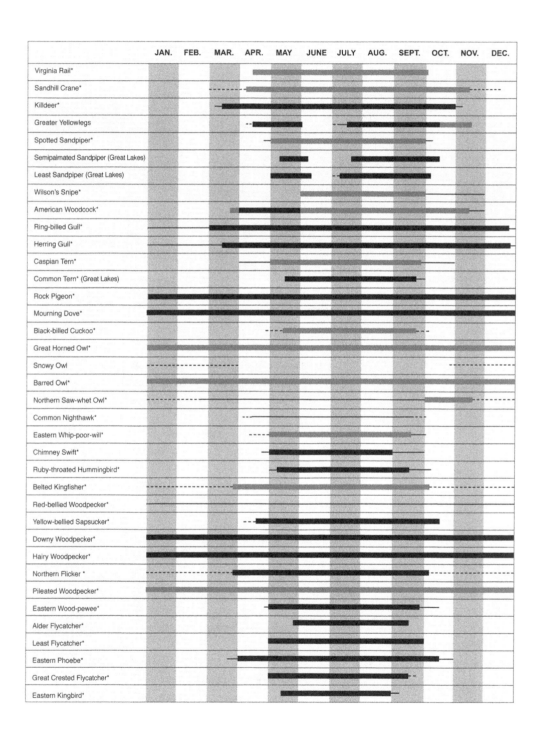

	JAN.	FEB.	MAR.	APR.	MAY	JUNE	JULY	AUG.	SEPT.	OCT.	NOV.	DEC.
Virginia Rail*												
Sandhill Crane*												
Killdeer*												
Greater Yellowlegs												
Spotted Sandpiper*												
Semipalmated Sandpiper (Great Lakes)												
Least Sandpiper (Great Lakes)												
Wilson's Snipe*												
American Woodcock*												
Ring-billed Gull*												
Herring Gull*												
Caspian Tern*												
Common Tern* (Great Lakes)												
Rock Pigeon*												
Mourning Dove*												
Black-billed Cuckoo*												
Great Horned Owl*												
Snowy Owl												
Barred Owl*												
Northern Saw-whet Owl*												
Common Nighthawk*												
Eastern Whip-poor-will*												
Chimney Swift*												
Ruby-throated Hummingbird*												
Belted Kingfisher*												
Red-bellied Woodpecker*												
Yellow-bellied Sapsucker*												
Downy Woodpecker*												
Hairy Woodpecker*												
Northern Flicker *												
Pileated Woodpecker*												
Eastern Wood-pewee*												
Alder Flycatcher*												
Least Flycatcher*												
Eastern Phoebe*												
Great Crested Flycatcher*												
Eastern Kingbird*												

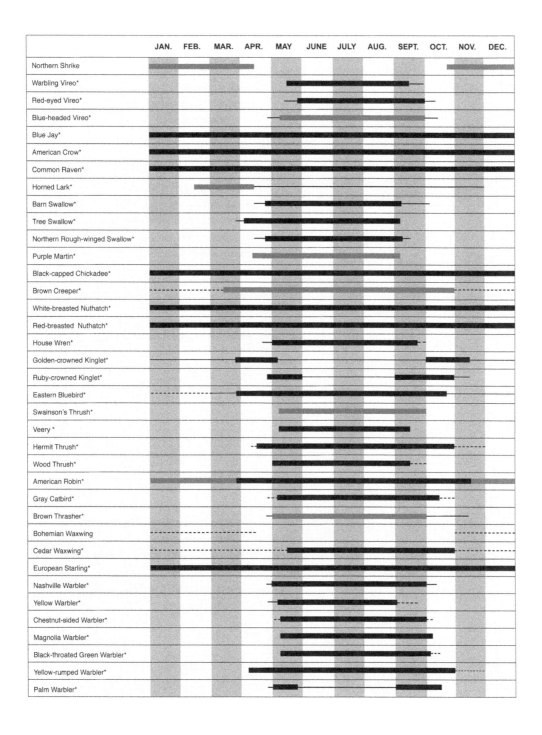

	JAN.	FEB.	MAR.	APR.	MAY	JUNE	JULY	AUG.	SEPT.	OCT.	NOV.	DEC.
Northern Shrike												
Warbling Vireo*												
Red-eyed Vireo*												
Blue-headed Vireo*												
Blue Jay*												
American Crow*												
Common Raven*												
Horned Lark*												
Barn Swallow*												
Tree Swallow*												
Northern Rough-winged Swallow*												
Purple Martin*												
Black-capped Chickadee*												
Brown Creeper*												
White-breasted Nuthatch*												
Red-breasted Nuthatch*												
House Wren*												
Golden-crowned Kinglet*												
Ruby-crowned Kinglet*												
Eastern Bluebird*												
Swainson's Thrush*												
Veery *												
Hermit Thrush*												
Wood Thrush*												
American Robin*												
Gray Catbird*												
Brown Thrasher*												
Bohemian Waxwing												
Cedar Waxwing*												
European Starling*												
Nashville Warbler*												
Yellow Warbler*												
Chestnut-sided Warbler*												
Magnolia Warbler*												
Black-throated Green Warbler*												
Yellow-rumped Warbler*												
Palm Warbler*												

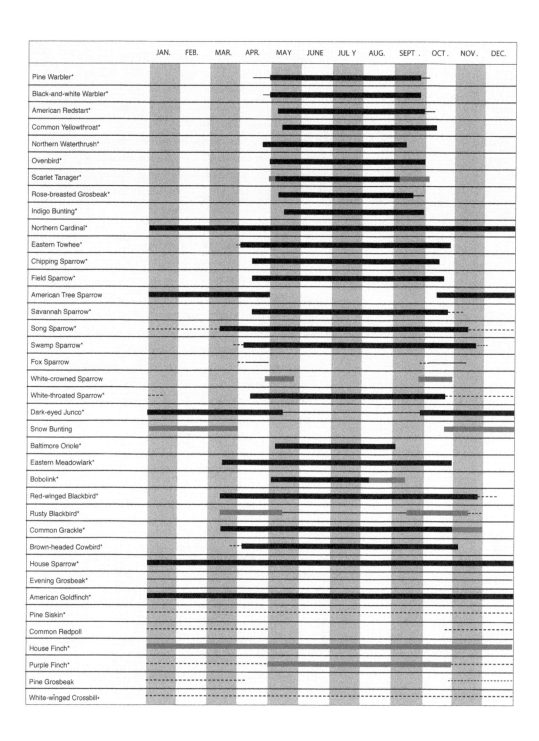

	JAN.	FEB.	MAR.	APR.	MAY	JUNE	JUL Y	AUG.	SEPT .	OCT .	NOV .	DEC.
Pine Warbler*												
Black-and-white Warbler*												
American Redstart*												
Common Yellowthroat*												
Northern Waterthrush*												
Ovenbird*												
Scarlet Tanager*												
Rose-breasted Grosbeak*												
Indigo Bunting*												
Northern Cardinal*												
Eastern Towhee*												
Chipping Sparrow*												
Field Sparrow*												
American Tree Sparrow												
Savannah Sparrow*												
Song Sparrow*												
Swamp Sparrow*												
Fox Sparrow												
White-crowned Sparrow												
White-throated Sparrow*												
Dark-eyed Junco*												
Snow Bunting												
Baltimore Oriole*												
Eastern Meadowlark*												
Bobolink*												
Red-winged Blackbird*												
Rusty Blackbird*												
Common Grackle*												
Brown-headed Cowbird*												
House Sparrow*												
Evening Grosbeak*												
American Goldfinch*												
Pine Siskin*												
Common Redpoll												
House Finch*												
Purple Finch*												
Pine Grosbeak												
White-winged Crossbill*												

APPENDIX 2

Where to Go

As Doug Sadler says in *Our Heritage of Birds*, "Birds are where you find them." In other words, interesting sightings of our flora and fauna can be had just about anywhere — even in your own backyard. The following locations, however, are well-known in the naturalist community as offering a particularly rich assortment of flora and fauna. They are listed by region and by their proximity to one another. Please refer to the map on page 309, or for more detailed information, highlights, and a list of amenities, I recommend visiting each location's website. Several of the sites listed are nature reserves owned by Ontario Nature. They are all open to visitors.

GREY-BRUCE

1. **Bruce Trail:** Follows the Georgian Bay shoreline and the top of the Niagara Escarpment from Collingwood west to Owen Sound and north to Tobermory. *Highlights*: Most spectacular views in central and southern Ontario; diverse geological formations; fascinating flora, including rare and abundant ferns; huge diversity of breeding birds.

2. **Bruce Peninsula National Park:** Located south of Tobermory at the tip of the Bruce Peninsula. *Highlights*: White dolomite cliffs of the Niagara Escarpment and the aquamarine waters of Georgian Bay make for spectacular scenery; ancient cedars and northern nesting birds; Dorcas Bay (Singing Sands) on Lake Huron shoreline has wetlands (fens) that are home to many types of orchids and insectivorous plants, Atlantic coastal species, and Great Lake endemics; massasauga rattlesnake; dark skies for astronomy; excellent visitor centre.

3. **Fathom Five National Marine Park:** Made up of the islands and surrounding waters off the tip of the Bruce Peninsula at Tobermory. *Highlights*: Iconic rock formations of Flower Pot Island; abundant eastern garter snakes; rare ferns and orchids.

4. **Cabot Head:** Located on the northeast tip of the Bruce, take road from Dyer's Bay to Cabot Head lighthouse. *Highlights*: Designated as an Important Bird Area; concentrates migrating songbirds and water birds (e.g., Red-necked and Horned Grebes, loons, diving ducks) in spring; spring raptor migration passes over this area; home of the Bruce Peninsula Bird Observatory; spectacular scenery.

5. **Fishing Islands:** A string of islands that border the western shore of the peninsula from the north end of Sauble Beach to Stokes Bay. Access is by boat. *Highlights*: Large variety of colonial nesting birds; characteristic fen plants of the Bruce Peninsula.

6. **Black Creek Provincial Park:** From Highway 6 at Ferndale, take Burma Road West and turn right onto Stokes Bay Road. Travel north to Myles Bay Road. Turn left and then right onto Sandy Bay Road. *Highlights*: Nesting Olive-sided Flycatcher and wood warblers such as Northern Parula; Stokes Bay is a resting and feeding area for diving ducks.

7. **Petrel Point Nature Reserve (Ontario Nature):** From Highway 6, north of Wiarton, take Red Bay Road west and then Huron Road north to Petrel Point Road. *Highlights*: Rich mosaic of fen and cedar swamps containing abundant orchids, pitcher plants, sundews, grass-of-parnassus, tuberous Indian-plantain, etc.

8. **Georgian Bay Shoreline (Dyer's Bay — Cape Chin — Lion's Head — Hope Bay):** Follow 40 Hills Road from just north of Wiarton along Georgian Bay shoreline. (Road changes names several times.) Provincial nature reserves at Lion's Head and Hope Bay. *Highlights*: Access to the Georgian Bay shoreline and the spectacular cliffs and bluffs of the Niagara Escarpment; Bruce Trail provides walking access to many spectacular sites; diverse ferns.

9. **Ferndale Flats:** Made up of the pasturelands surrounding the hamlet of Ferndale on Highway 6. *Highlights*: Grassland birds, most notably western species such as Brewer's Blackbird and sometimes Western Meadowlark; flocks of Sandhill Cranes in fall; Snowy Owls and Rough-legged Hawks in winter.

10. **Sky, Boat, and Isaac Lakes:** Chain of three lakes located west of Wiarton. *Highlights*: Extensive breeding populations of wetland birds, including bitterns, Black Terns, rails, waterfowl, and Marsh Wrens.

11. **Chantry Island:** Located in Lake Huron off the town of Southampton on Highway 21 — access by boat. *Highlights*: A very significant bird sanctuary due to its diversity of colonial nesting birds, including Great Egret, Black-crowned Night-heron, and Great Black-backed Gull; large numbers of migrating diving ducks each spring and fall.

12. **MacGregor Point Provincial Park:** Located south of Port Elgin on Highway 21 on Lake Huron shoreline. Follow the signs. *Highlights*: Located on the "Huron Fringe," some of the richest biodiversity in Ontario with coastal dunes, bluffs, cobble shore, and a wide variety of wetland types and associated flora such as the rare dwarf lake iris; migrating birds; salamander viewing in early spring; Huron Fringe Birding Festival (late May–early June) offers numerous activities to learn about all aspects of nature.

13. **Inverhuron Provincial Park:** Travel west from Tiverton on Highway 21 between Port Elgin and Kincardine. Follow signs. *Highlights*: Sand dunes home to Great Lake specialties such as pitcher's thistle and dwarf sandcherry; great birding spring and fall; water birds concentrate in the waters off the park in the fall.

14. **Point Clark:** Take the Ashfield-Huron Line west from Amberley on Highway 21, south of Kincardine. At the lake, take Lake Range Road to Huron Road. Turn left onto Lighthouse Road and follow to the lighthouse. *Highlights*: Excellent location in the fall to see birds that would normally be offshore, such as waterfowl, rare gulls, and even jaegers; huge concentration of Red-throated Loons in April.

15. **Greenock Swamp:** 20,000 acres in size, the swamp extends north and south of Highway 9 between Kincardine and Walkerton. There are trails in the Schmidt/Bester Lakes area west of Dunkeld and Chepstow. *Highlights*: A rich assemblage of woodland flora and fauna, including diverse breeding birds like Cerulean Warblers.

16. **Saugeen River Canoe Route:** River flows from Hanover, through Walkerton and Paisley,

and terminates at Southampton on Lake Huron. *Highlights*: A popular canoe route, river banks contain some Carolinian trees such as sycamore and common hackberry.

17. **Kemble Mountain:** Northwest of Owen Sound. Follow Kemble Bay Road north from the hamlet of Kemble to the Kemble Mountain Management Area. *Highlights*: Upland hardwood forest; 20 species of ferns, including large concentrations of Hart's-tongue fern; trails with spectacular views.

18. **Owen Sound Bay:** Several public areas are located along the shoreline of the bay, such as Indian Creek and Ainslie Woods conservation areas. *Highlights*: Concentrations of water birds from September to May.

19. **Inglis Falls Conservation Area:** Take 2nd Avenue south from downtown Owen Sound to County Road 5. Turn left and follow to Inglis Falls Road. *Highlights*: Mature mixed woodland with wide diversity of ferns, including Hart's-tongue fern; spectacular 18-metre waterfall; access to the Bruce Trail.

20. **Beaver Valley:** Extends south of Meaford and Thornbury. A number of protected areas on both sides of the valley, including Epping Lookout, Old Baldy, and Eugenia Falls. *Highlights*: Scenic lookouts across the valley from the cliffs of the Niagara Escarpment; scenic waterfalls; cliff-loving ferns.

21. **Kinghurst Forest Nature Reserve (Ontario Nature):** Take Highway 6 north from Durham to Dornoch and turn west onto County Road 25. Go about eight kilometres to Concession Road 6 and turn north. Just past a side road on left is a driveway leading to a white house. ONTARIO NATURE sign marks entrance to main trail. *Highlights*: Remarkable old-growth maple-beech forest with many trees up to 300 years old; spectacular wildflowers in spring.

22. **Feversham Gorge:** On County Road 2 at Feversham, park in Madeleine Graydon Conservation Area. Follow marked trail to gorge. *Highlights*: A botanist's paradise, thanks to abundant and diverse mosses, ferns, and liverworts; Brook Trout.

LAKE SIMCOE REGION (DUFFERIN, YORK, AND SIMCOE COUNTIES)

23. **Awenda Provincial Park:** Located on Georgian Bay, approximately 20 minutes northwest of Penetanguishene. *Highlights*: mixed forest; spring wildflowers, including painted trillium; mix of northern and southern nesting birds; extensive trail network.

24. **Tiny Marsh Provincial Wildlife Area:** From Elmvale, take County Road 27 north to Flos Road. Turn left and go 3.7 kilometres to entrance. *Highlights*: Abundant waterfowl, marsh birds, shorebirds.

25. **Matchedash Bay:** approximately five kilometres north of Coldwater and immediately east of Waubaushene. *Highlights*: Important Bird Area; diverse breeding marshland birds, including significant numbers of Least Bittern and Black Terns; numerous and diverse waterfowl in spring.

26. **Minesing Wetlands Conservation Area:** Highway 400 north to County Road 90, then west toward Angus. Various access points to the swamp north of Highway 90 include McKinnon,

Sunnidale, and George Johnston Roads. *Highlights*: World-class wetland with numerous water-fowl species and shorebirds during migration; raptors.

27. Mono Cliffs Provincial Park: From Orangeville, take Highway 10 north to County Road 8 at Camilla. Turn right and continue to park entrance just north of Mono Centre. *Highlights*: Diverse Niagara Escarpment habitats supporting a profusion of ferns and old-growth white cedars; Jefferson complex salamanders.

28. Cawthra Mulock Nature Reserve: From Newmarket, go north on Dufferin Street past Millers Side Road. West entrance to reserve is on the right, across from 18580 Dufferin Street; east entrance at 18462 Bathurst Street. *Highlights*: A large tract of mature hardwood and mixed forests in a highly developed part of Ontario includes enormous beech trees; a wetland dominated by birch and tamarack; two creeks; nesting Red-tailed Hawks and Great Horned Owls.

YORK-DURHAM

29. Nonquon Sewage Lagoons: Located on Scugog Line 8, east off Highway 12, north of Port Perry. *Highlights*: Diverse and sometimes abundant migrating shorebirds in spring through fall; Black Terns. *Note*: A five-dollar permit is necessary. Obtain at Durham Region Transfer Site at 1623 Reach Road, Port Perry.

KAWARTHA LAKES-PETERBOROUGH-NORTHUMBERLAND

30. Carden Alvar: Located northeast of Beaverton. From Kirkfield, take County Road 6 north. Explore concessions such as Wylie Road and Alvar Road. *Highlights*: One of the best late spring and early summer birding destinations in central Ontario, especially for grassland birds (Loggerhead Shrike, Upland Sandpiper, Eastern Bluebird, Sedge Wren, rare sparrows, etc.); marsh birds, sometimes including Yellow Rail; unique alvar plant communities (prairie smoke, Indian paintbrush); Carden Nature Festival in early June with numerous activities to learn about all aspects of nature.

31. Altberg Wildlife Sanctuary Nature Reserve (Ontario Nature): From Norland on Highway 35, take County Road 45 east for seven kilometres. The property is on the south side of the road at address marker 4164. *Highlights*: 470 hectares straddling the contact between the granite rocks of the Canadian Shield and the limestone of the Great Lakes–St. Lawrence Lowlands; great diversity of habitat types, breeding birds (e.g., Red-shouldered Hawk), and flora (e.g., showy lady's slipper).

32. Ken Reid Conservation Area: From junction of Highways 7 and 35, go five kilometres north on Highway 35. Turn right on Kenrei Park Road and continue three kilometres. *Highlights*: Huge marsh with boardwalks; forest, meadows, and marshes; high density of active Osprey nests.

33. Fleetwood Creek Natural Area: From Pontypool, drive north on Highway 35. Turn right onto Ballyduff Road and continue for 2.9 kilometres. *Highlights*: 380-hectare property located within the Oak Ridges Moraine; trails through mature lowland forests, meadows, steep valleys; diverse flora; interesting geological formations; impressive fall foliage.

34. Ganaraska Forest: From Port Hope, take County Road 28 north to County Road 9. Turn left and continue to Northumberland–Durham boundary. Turn right and go three kilometres to Ganaraska Forest Centre. *Highlights*: Located on the Oak Ridges Moraine; southern Ontario's largest forest; hundreds of kilometres of trails; good selection of woodland birds, including Hooded Warbler; new Ganaraska Forest Centre, which houses the Oak Ridges Moraine Information Centre, outdoor education facilities, etc.

35. Little Lake: From south end of downtown Peterborough on George Street, go east along shore to well-treed cemetery. *Highlights*: Early spring and late fall waterfowl, especially diving ducks and grebes; Glaucous, Iceland, and Great Black-backed Gulls possible in late fall and winter.

36. River Road and Lakefield Sewage Lagoon: Also called Peterborough County Road 32. Located on east bank of Otonabee River between Peterborough and Lakefield. Lagoon located on County Road 33, just east of County Road 32, on south edge of Lakefield. Please do not block gate. *Highlights*: River has diving ducks during migration and in winter; migrating swallows in spring; Bald Eagles possible in winter; lagoon excellent for diverse duck species during migration; Black Terns in summer.

37. Mark S. Burnham Provincial Park: Located on north side of Highway 7, just east of junction with Television Road on east side of Peterborough. *Highlights*: Remnant stand of very large native trees (e.g., sugar maple, hemlock); rich and diverse flora, including ferns, sedges, spring wildflowers.

38. Petroglyphs Provincial Park: Follow Highway 28 north from Burleigh Falls to just past Woodview. Turn right on Northey's Bay Road and follow for about 11 kilometres. *Highlights*: Situated on southern edge of Canadian Shield; extensive trail system; diverse flora, including large stands of red and white pine; abundant white-tailed deer; birds of interest often include eagles and crossbills in winter; five-lined skinks are fairly common; diverse butterflies.

39. Kawartha Highlands Provincial Park: Located north of Buckhorn Lake between County Road 507 and Highway 28. Access from Anstruther Lake Road at Apsley. *Highlights*: A huge park with vast rock barrens and strong wilderness qualities; high-quality bog, fen communities, alvar and Atlantic coastal plain plant communities, all with interesting flora; mature forest stands; high concentrations of some species at risk, including Whip-poor-will and Common Nighthawk; dark skies for astronomy.

40. Sandy Lake Road: Follow County Road 46 north from Havelock. Turn right about six kilometres north of Oak Lake onto Sandy Lake Road. *Highlights*: Diverse butterflies, including uncommon skippers (e.g., mulberry wing, broad-winged) along the edge of the sedge marshes.

41. Peter's Woods Provincial Nature Reserve: From Centreton, travel north on McDonald Road exactly five kilometres to entrance on east side. *Highlights*: Magnificent old-growth forest with diverse ferns (e.g., maidenhair), orchids (showy orchis), spring wildflowers, and birds (southern species possible).

42. Goodrich-Loomis Conservation Area: From Brighton, go 12.6 kilometres north on Highway 30 to junction with Goodrich Road. Turn left and follow. *Highlights*: A wide variety of ecological communities; one of eastern Ontario's finest trout streams; oak savannah;

wetland; mature mixed forest; well-established bluebird population. A four-hectare prairie site is being restored.

43. Presqu'ile Provincial Park: Located south of Brighton on Lake Ontario. Follow signs. *Highlights*: Ten-kilometre-long peninsula jutting into Lake Ontario; migrant trap for songbirds; water birds and shorebirds migrate through in large numbers; interesting plant communities of over 700 species, including false dragonhead, grass-of-parnassus, and Kalm's lobelia; staging area for migrant monarch butterflies. Amenities include trails, marsh boardwalk, bird sightings board, visitor centre; special event weekends throughout year, including Waterfowl Viewing Weekend in March.

MUSKOKA

44. Arrowhead Provincial Park: Located six kilometres north of Huntsville on Highway 11. *Highlights*: Extensive hardwood forests with beautiful fall colours; interesting flora, including cardinal flower; excellent trail system.

45. Hardy Lake Provincial Park: Located 6.5 kilometres east of Bala on County Road 169. *Highlights*: Rich community of Atlantic coastal plain flora, including spike rush, golden hedge hyssop, and Virginia meadow beauty; excellent trail system.

46. Torrance Barrens Conservation Reserve: From Torrance on County Road 169, take County Road 13 south for five kilometres. Park near Highland Pond. *Highlights*: Bedrock barrens treed with red and white oak and substantial component of white and red pine; interesting flora and reptiles (e.g., skinks, massasauga rattlesnakes).

47. Six Mile Lake Provincial Park: Off Highway 400, north of Port Severn. *Highlights*: Rock barrens; interesting flora, including yellow-eyed grass and smartweed; abundant birdlife (e.g., Yellow-throated Vireo); five-lined skinks.

48. Georgian Bay Islands National Park: From Honey Harbour, take canoe or water taxi. *Highlights*: Sixty islands in the park; abundant shoreline and wetland habitat with rich flora (e.g., triangle grape fern, southern twayblade); diverse amphibians (two-lined salamander) and reptiles of 35 species (e.g., hog-nosed snake, massasauga rattlesnake, five-lined skink); Prairie Warblers inhabit open areas; diverse butterflies.

PARRY SOUND

49. Killbear Provincial Park: Located north of Parry Sound on County Road 559. *Highlights*: Spectacular Georgian Bay lakeshore scenery; good mammal and bird viewing opportunities; rare reptiles, including map turtle and massasauga rattlesnakes.

50. The Massasauga Provincial Park: Accessible only by water. Canoeists can enter via Three Legged Lake. For trips into Georgian Bay, start at Pete's Place on Blackstone Harbour. *Highlights*: Hundreds of windswept islands; high cliffs and huge granite outcrops; diverse turtle and snake (e.g., northern ribbon snake) fauna; sanctuary for massasauga rattlesnake; nesting Prairie Warblers; Atlantic coastal plain flora (e.g., Virginia meadow-beauty); sphagnum bogs.

HALIBURTON

51. Algonquin Provincial Park: From Bancroft, take Highways 62 and 127 north to Highway 60. Turn left and follow Highway 60 through southern sector of park (most of which is in Nipissing District). *Highlights*: Transition zone between the southern hardwood and boreal evergreen forests with diverse flora and fauna; boreal birds such as Gray Jay, Spruce Grouse, and Boreal Chickadee at locations such as Spruce Bog Trail; abundant winter finches (e.g., crossbills) some years; self-guiding interpretive trails; dark skies for astronomy; excellent visitor centre with active feeders.

52. Silent Lake Provincial Park: Located on Highway 28 between Apsley and Bancroft. *Highlights*: Diverse habitats, including mixed medium-aged forests, sphagnum bogs, beaver meadows; valleys support 25 species of ferns (e.g., silvery glade fern); four-toed salamander.

53. Bentshoe Lake: On north side of Highway 118, 18 kilometres west of Carnarvon. *Highlights*: Rich assemblage of aquatic species of Atlantic coastal plain flora, especially in northeastern portion of lake.

HASTINGS AND PRINCE EDWARD

54. The Gut Conservation Area: Located east of Apsley. From junction of County Roads 504 and 46 at Lasswade, go 6.5 kilometres east. *Highlights*: Walking trials; a 30-metre-high gorge on the scenic Crowe River; north woods plants, including trailing arbutus; good birding.

55. Massassauga Point Conservation Area: From Belleville, go south on Highway 62. Turn left at County Road 28 past bridge and left again on Massassauga Road. Continue eight kilometres to area. *Highlights*: Meadow home to alvar plants like prairie smoke and sideoats grama grass; ash and hickory forest; good birding.

56. Lake on the Mountain Provincial Park: Eight kilometres east of Picton on Highway 33. *Highlights*: turquoise lake on top of a 60-metre escarpment above the waters of Lake Ontario.

57. Prince Edward Point National Wildlife Area: From Picton, take County Road 8 east for 2.5 kilometres, then County Road 17 southeast for 6.5 kilometres. Turn left onto County Road 6 and follow for one kilometre to County Road 13. Follow for 23.6 kilometres to destination. *Highlights*: Concentrates migrants in spring and fall, including songbirds, waterfowl, hawks, owls, and monarch butterflies; overwintering area for waterfowl in winter; home of the Prince Edward Point Bird Observatory; Saw-whet Owl banding.

58. Sandbanks Provincial Park: Located 14 kilometres southwest of Picton on County Road 12. *Highlights*: active sand dune system with beach, dunes (up to 25 metres), pannes (damp grassland between dunes), and forested dunes; diverse plants, including butterfly milkweed, hoary puccoon, and purple gerardia; migrating shorebirds in spring through fall.

LAND O' LAKES (LENNOX AND ADDINGTON, FRONTENAC)

59. Bon Echo Provincial Park: Located about 30 kilometres north of Kaladar on Highway 41. *Highlights*: Massive 125-metre-high granite-gneiss cliff; rock outcrops; excellent fall colour; ridges of mature forest (e.g., yellow birch, American beech); numerous unusual plants (e.g., smooth cliffbrake, green alder); rich mix of northern and southern songbirds and mammals; dark skies for astronomy; numerous trails and scenic lookouts.

60. Frontenac Provincial Park: Located north of Sydenham, off County Road 19. *Highlights*: Extensive tracts of mature hardwood forest; fern-filled valleys; abundant beaver ponds; mixture of northern and southern landscapes with many diverse species (e.g., five-lined skink, black rat snake, Hooker's orchid).

61. Opinicon Road Area: From Highway 401, take County Road 10 (exit 617) about 25 kilometres north to just past Raymonds Corners. Take Opinicon Road on your right. *Highlights*: Excellent birding, including Yellow-throated Vireo, cuckoos, and rare warblers such as Cerulean and Golden-winged; map turtles in this area (e.g., Chaffey's Locks).

62. Little Cataraqui Creek Conservation Area: From Kingston go north on Division Street (County Road 10) 1.5 kilometres past Highway 401. *Highlights*: Diverse habitats, including mixed woodlands, sugar maple bush, ponds; many songbirds and interesting plants along trails; Outdoor Education Centre.

63. Lemoine Point Conservation Area: Two entrances; one off Highway 33 at Collin's Bay and one off Front Road (County Road 1) past the Kingston airport. *Highlights*: Abundant spring wildflowers in deciduous woods with mature hardwoods; many species of resident and migratory birds.

64. Wolfe Island: In downtown Kingston, take ferry from Ontario Street near foot of Queen Street. *Highlights*: Similar to Amherst Island; birding is excellent from concession roads and roads along the shore.

65. Amherst Island: Take County Road 4 south from Highway 401 to Millhaven. Take ferry to island. *Highlights*: Extensively used by wintering hawks (e.g., Rough-legged, Northern Harrier) and at least five owl species, including Snowy, Short-eared, Long-eared, and Northern Saw-whet (mostly in Owl Woods at east end of island); large rafts of diving ducks during migration and in winter; diverse shorebirds in late summer. *Note*: Please observe all posted rules to reduce stress on owls.

OTTAWA VALLEY (RENFREW COUNTY)

66. Bonnechere, Bonnechere River, and Foy provincial parks: All three parks are located close to town of Bonnechere on Round Lake, northeast of Barry's Bay. *Highlights*: Bonnechere River includes the shorelines of the Little Bonnechere River from Round Lake to Algonquin Provincial Park, where forested uplands (e.g., hemlock, pine, maple) tower 300 metres above the valley floor; Foy has remnants of Renfrew County's original hemlock-white pine and hemlock-white cedar forests; Bonnechere covers 162 hectares of forest and wetland; rich flora and fauna throughout this area.

67. Shaw Woods: Go 13 kilometres north from Eganville on Highway 41. Woods are two kilometres south of Highway 41 on Bulger Road. *Highlights*: Towering stands of old-growth maple, beech, and hemlock up to 35 metres in height; variety of wetland types, including ancient white cedar swamp; outdoor education centre.

68. Centennial Lake Provincial Nature Reserve: Black Donald Island is of special interest to naturalists. Accessible by boat or canoe. From Calabogie, take County Road 508 west to public boat launch on north shore of Black Donald Lake, two kilometres northeast of island. *Highlights*: Rare plants, including communities of purple-stemmed cliffbrake, mountain woodsia, New Jersey tea, long-stemmed waterwort; Bald Eagles year-round at Mountain Chute dam.

69. Stewartville Swamp Nature Reserve (Ontario Nature): From Arnprior, follow Russett Drive (County Road 45) west through Stewartville and past the junction of Highway 63. Swamp is on right side through the cedar woods, about one kilometre past Stewartville. No formal trail. *Highlights*: Abundant orchids of 18 species (including close to 500 dwarf rattlesnake plantains) grow beneath big cedars; many kinds of ferns.

The spectacular showy lady's slipper is one of many orchids to be found at the Stewartville Swamp Nature Reserve.
Kim Caldwell

70. **Ottawa River:** Locations for spotting water birds include mouths of the Bonnechere (east of Castleford) and Madawaska (Arnprior) rivers. *Highlights:* A major migration corridor for water birds in spring (mid-April) and fall (late September to late October) linking directly to James Bay; superb scenery between Mattawa and Deep River; Bald Eagles along some stretches.

71. **Westmeath Provincial Park:** Follow Ottawa River 16.5 kilometres east from Pembroke along Highway 148. *Highlights:* rare reptiles such as map turtle; exceptional flora (e.g., blue-beech, Atlantic coastal plain and dune plant species); shorebirds in fall.

Rideau Lakes (Lanark, Leeds, and Grenville)

72. **Purdon Conservation Area:** From Perth, go north on County Road 511 through Lanark Village. Continue ten kilometres and turn west onto County Road 8. Follow through Watson's Corners and continue ten kilometres. Turn north on Concession Road 8. Follow signs to entrance. *Highlights:* Huge showy lady's slipper orchid display in mid-June to early July; forest trail loop.

73. **Foley Mountain Conservation Area:** Go 0.4 kilometres north of Westport on County Road 10 and follow signs. *Highlights:* Breathtaking lookouts; 60-metre-high granite rock outcrops; oak and pine forests; good numbers of breeding warblers.

74. **Charleston Lake Provincial Park:** From Highway 401, take exit 659 to Lansdowne. Located 20 kilometres north off County Road 3. *Highlights:* Sits on the Frontenac Axis, a southerly extension of the Precambrian shield that divides the St. Lawrence Lowlands in two and gives rise to the Thousand Islands and the Adirondack Mountains of New York; rich flora and fauna with both northern and southern elements; lush wetlands, including black spruce bogs; scenic lookout at Blue Mountain; rare reptiles such as black rat snakes; dark skies for astronomy.

75. **Lost Bay Nature Reserve (Ontario Nature):** Going east on Highway 401 past Gananoque, take exit 648 to County Road 2. Go east to Kyes Road. Travel north on Kyes Road (name changes) past Sand Bay Corner. Take Lost Bay Lane cottage road to reserve. *Highlights:* Part of the Algonquin to Adirondack Connection because it is located within the Frontenac Axis; provincially significant wetlands and mature forest; wide variety of birds, including Red-shouldered Hawk.

76. **St. Lawrence Islands National Park:** For mainland portion of park, take exit 675 from Highway 401 and go 3.5 kilometres south to Mallorytown Landing. Access to islands is by boat. *Highlights:* Resident population of the threatened black rat snake; Canada's largest stand of pitch pine (Hill Island); eastern species such as red spruce and gray birch; southern species such as rue anemone, chinquapin oak; abundant birdlife, including wintering Bald Eagles between Ivy Lea and Brockville.

77. **Ferguson Forest Centre:** From County Road 43 in Kemptville, go north on County Road 44. Centre is on right. *Highlights:* Diverse habitat, including extensive mixed woodlands, hardwood forest, and extensive marsh; good diversity of wildlife; interesting flora, including painted trillium in spring; large trail network.

78. Oxford Mills: From Highway 416, take exit 34 to Kemptville. Take County Road 44 south. Turn right on County Road 18 and follow southwest to Oxford Mills. *Highlights*: Mudpuppy salamanders; "Mudpuppy Night in Oxford Mills" takes place every Friday night at 8:00 p.m. from Thanksgiving to early spring.

Ottawa-Carleton Region

79. Shirley's Bay: Take exit 134 from Highway 417, go 1.5 kilometres north on Moodie Drive, then 2.5 kilometres west on Carling Avenue, then two kilometres north on Range Road to parking area. Permission is needed to enter the dyke area. Call 613-991-5740. *Highlights*: Abundant ducks, shorebirds, and songbirds that migrate along Ottawa River; earthen berm provides excellent views of birds; alvar-shrub prairie (e.g., shrubby St. John's wort) at Innis Point to the north of the bay.

80. Andrew Haydon Park: Located along shore of Britannia Bay. From Ottawa, continue west on Carling Avenue. Watch for a small parkland with play structures. *Highlights*: Excellent fall birding for shorebirds, gulls, and water birds, including grebes and sea ducks; extensive sand flats.

81. Britannia Conservation Area: Take Pinecrest Road (exit 129) off Highway 417. Go north on Pinecrest, then turn right onto Richmond Road. Past Carling Avenue, turn left onto Poulin Avenue and then right onto Britannia Road. Britannia Road and Cassels Street form the BCA's west and north boundaries. *Highlights*: Great diversity of habitats; abundant songbirds during migration; gulls in spring and fall; Arctic Terns possible in late May from Britannia Point.

82. Mer Bleue Conservation Area: Take exit 104 on Highway 417 east of Ottawa. Go north on Regional Road 27 for four kilometres, then turn right on Borthwick Ridge Road. Boardwalk at end. *Highlights*: Huge northern bog habitat and related flora; red maple swamps with abundant ferns; boreal nesting birds such as Palm Warbler and Lincoln's Sparrow.

83. Low-lying Farmlands: Try Boundary Road 31 north of Carlsbad Springs and Regional Road 49 east of Richmond. *Highlights*: Large flocks of waterfowl in flooded fields in early spring, including Snow Geese and many ducks; grassland birds; Snowy Owls in winter.

84. South March Highlands: Located west of Kanata Lake area of Kanata along Terry Fox Drive, Old Carp Road, and Huntmar Drive. *Highlights*: Outcrops of Precambrian bedrock; rich sugar maple hardwoods with uncommon ferns (e.g., Goldie's wood fern) and spectacular spring wildflowers; mix of northern and southern flora and fauna.

85. The Burnt Lands: From Almonte, go five kilometres east on County Road 49. South side is public land. *Highlights*: Extensive alvar habitat (areas of thin soil over flat-lying limestone bedrock) with many unique plants such as prairie dropseed, balsam ragwort, upland white goldenrod; spectacular spring floral display in white cedar glades (e.g., yellow lady's slipper, wood lily).

Stormont-Dundas-Glengarry and Prescott and Russell

86. Upper Canada Migratory Bird Sanctuary: From Highway 401 (exit 758), go south to County Road 2. Turn right. The sanctuary is between Riverside Heights and Ingleside. *Highlights*:

Great in early spring and late fall for abundant waterfowl, including Snow Geese; good general birding year-round; interpretive centre.

87. Moses-Saunders Power Dam: Take Highway 401 exit to Brookdale Avenue in Cornwall. Follow 2.3 kilometres to Toll Gate Road. Viewing is best from Hawkin's Point on American side. *Highlights*: Area below dam has excellent variety of gull species in fall and winter (e.g., Glaucous, Iceland, Thayer's) and waterfowl (e.g., mergansers, goldeneyes, Long-tailed Ducks); loons, grebes, and scoters above dam.

88. Cooper Marsh Conservation Area: From Highway 401 (exit 814), go one kilometre south to South Lancaster. Turn right on County Road 2 and follow for three kilometres to area. *Highlights*: Large St. Lawrence River wetland managed for waterfowl production; magnificent marsh boardwalk and views of waterfowl.

89. Alfred Bog: From Ottawa, take Highway 417 east to County Roads 174/17 and continue east. Approximately two kilometres before town of Alfred, turn right on Station Road and continue to end of road. *Highlights*: Superb peat bog ecosystem and related flora, such as Labrador tea and rare orchids; in spring, Lincoln's and Clay-coloured Sparrows along board-walk; Sandhill Cranes in fall; resident moose herd.

90. Alfred Sewage Lagoons: From just east of Alfred on County Road 17, go south on Peat Moss Road past Rang Saint Jean (Concession Road 7). Lagoons are on left. *Highlights*: abundant shorebirds (e.g., Wilson's Phalarope, Whimbrel) and water birds (e.g., Ruddy Duck, American Coot, Virginia Rail, Sora) can be found in season; birding tower offers great views.

WHERE TO GO:
NATURE-VIEWING AREAS IN CENTRAL AND EASTERN ONTARIO

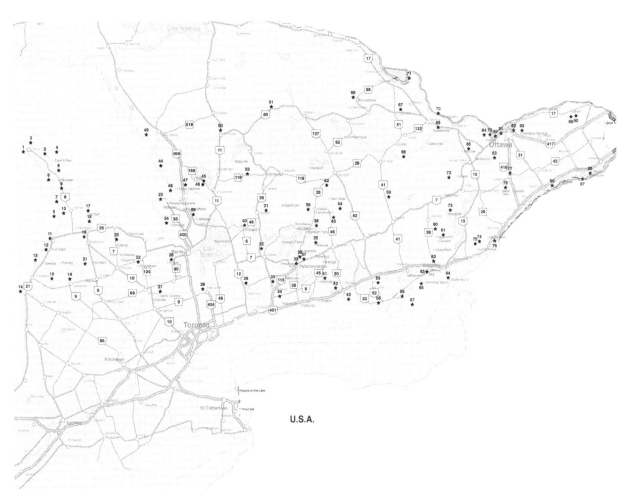

This map shows the location of some of the best places in central and eastern Ontario to view nature.

Jeff Macklin

NOTES

Introduction

1. Mike Oldham, botanist, Natural Heritage Information Centre, email, October 16, 2011.

2. Bernd Heinrich, *A Year in the Maine Woods* (Don Mills, ON: Addison-Wesley, 1994), 15.

3. Frontenac Arch Biosphere Reserve, "Why the Frontenac Arch is a UNESCO World Biosphere Reserve," *www.fabr.ca/Mandate%20and%20Statement.htm*.

4. Union of Concerned Scientists, "Climate Change and Ontario," *www.ucsusa.org/greatlakes/ glregionont.html*.

5. Some of the events and dates were taken from: Michael W.P. Runtz, *The Explorer's Guide to Algonquin Park* (Toronto: Stoddart Publishing, 1993), 9.

January

1. Marc Johnson, " Great Gray Owl," 298–99, in M.D. Cadman, D.A. Sutherland, G.G. Beck, D. Lepage, and A.R. Couturier, eds, *Atlas of the Breeding Birds of Ontario, 2001–2005* (Toronto: Bird Studies Canada, Environment Canada, Ontario Field Ornithologists, Ontario Ministry of Natural Resources, and Ontario Nature, 2007).

2. Project FeederWatch 2010–11, "Top 25 Feeder Birds 2010–2011 Ontario," *www.bsc-eoc.org/ volunteer/pfw/index.jsp?lang=EN&targetpg=index*.

3. Interview with Brock Fenton, "Scientists Say Bat Disease Likely to Spread," National Public Radio, *www.npr.org/templates/story/story.php?storyId=112035629*.

4. Jeffrey M. Lorch and others, "Experimental Infection of Bats with *Geomyces destructans* Causes White-nose Syndrome," *Nature* (2011) doi: 10.1038/nature10590. Published online October 26, 2011, *www.nature.com/nature/journal/vaop/ncurrent/full/nature10590.html*.

5. Rick Rosatte, Senior Research Scientist, MNR and Trent University, email, August 8, 2011.

6. National Climate Data and Information Archive, "Canadian Climate Normals or Averages 1971–2000," *www.climate.weatheroffice.gc.ca/climate_normals/index_e.html*.

7. Keith C. Heidorn, *And Now ... the Weather: With the Weather Doctor* (Calgary: Fifth House, 2005), 32.

8. Sunrise Sunset, "Custom Sunrise Sunset Calendar," *www.sunrisesunset.com*.

February

1. George K. Peck, "Common Raven," 384–85, in M.D. Cadman, D.A. Sutherland, G.G. Beck, D. Lepage, and A.R. Couturier, eds, *Atlas of the Breeding Birds of Ontario, 2001–2005*.

2. Jennifer R. Foote and others, "Black-capped Chickadee (*Poecile atricapillus*)," The Birds of North America Online, *bna.birds.cornell.edu/bna/species/039*.

3. A. Leopold, and S.E. Jones, 1947, "A Phenological Record for Sauk and Dane Counties, Wisconsin, 1935–1945," *Ecological Monographs 17*, 86.

4. Steve Lawrence, "Climate Change Putting Lake Trout at Risk," Haliburton Highlands Outdoors Association, *www.hhoa.on.ca/education.html*.

5. National Climate Data and Information Archive, "Canadian Climate Normals or Averages 1971–2000," *www.climate.weatheroffice.gc.ca/climate_normals/index_e.html*.

6. Sunrise Sunset, "Custom Sunrise Sunset Calendar," *www.sunrisesunset.com*.

March

1. Bernd Heinrich, *A Year in the Maine Woods* (Don Mills, ON: Addison-Wesley, 1994), 15.

2. National Climate Data and Information Archive, "Canadian Climate Normals or Averages 1971–2000," *www.climate.weatheroffice.gc.ca/climate_normals/index_e.html*.

3. Sunrise Sunset, "Custom Sunrise Sunset Calendar," *www.sunrisesunset.com*.

April

1. Peter Blancher, Michael D. Cadman, Bruce A. Pond, Andrew R. Couturier, Erica H. Dunn, Charles M. Francis, and Robert S. Rempel, "Changes in Bird Distributions Between Atlases," 42, in M.D. Cadman, D.A. Sutherland, G.G. Beck, D. Lepage, and A.R. Couturier, eds, *Atlas of the Breeding Birds of Ontario, 2001–2005*.

2. Don Sutherland, zoologist, Natural Heritage Information Centre, email, May 13, 2011.

3. Ken Taggart (compiler), "Kawartha Lakes Northern Pike Invasion Project (2009–Present)," Muskies Canada, *www.muskiescanada.ca/club_information/initiatives.php*.

4. Chantal Braganza, "Finally, a Suspect in Bee Decline," *www.thestar.com/news/insight/article/754993* (accessed September 12, 2011).

5. Lorraine Johnson, "Save the Bees: How to Attract Pollinators to Your Yard," *Canadian Gardening*, *www.canadiangardening.com/how-to/wildlife*.

6. National Climate Data and Information Archive, "Canadian Climate Normals or Averages 1971–2000," *www.climate.weatheroffice.gc.ca/climate_normals/index_e.html*.

7. Sunrise Sunset, "Custom Sunrise Sunset Calendar," *www.sunrisesunset.com*.

May

1. Cindy E.J. Cartwright, "Ruby-throated Hummingbird," 316–17, in M.D. Cadman, D.A. Sutherland, G.G. Beck, D. Lepage, and A.R. Couturier, eds, *Atlas of the Breeding Birds of Ontario, 2001–2005*.

2. Peter Blancher, Michael D. Cadman, Bruce A. Pond, Andrew R. Couturier, Erica H. Dunn, Charles M. Francis, and Robert S. Rempel, "Changes in Bird Distributions between Atlases," 42, in M.D. Cadman, D.A. Sutherland, G.G. Beck, D. Lepage, and A.R. Couturier, eds, *Atlas of the Breeding Birds of Ontario, 2001–2005*.

3. Robin Follette, "Black Fly Info," Maine Nature Notes, *mainenature.org/black-fly-info*.

4. The Regional Municipality of Halton: Public Health, "West Nile Virus and Mosquitoes," *www.halton.ca/cms/one.aspx?portalId=8310&pageId=10023*.

5. Jerry Ball, phone call, October 13, 2011.

6. Mark D. Schwartz, R. Ahas, and A. Aasa, 2006, "Onset of Spring Starting Earlier Across the Northern Hemisphere," *Global Change Biology*, 12, 343–51.

7. *Dukenvironment Magazine*, Fall 2006, "Global Warming May Increase Poison Ivy's Itch," *www.nicholas.duke.edu/DEarchives/f06/log-ivy.html*.

8. Richard Alleyne, "Feeling Stressed? Then Go Mow the Lawn," *www.telegraph.co.uk/science/6094786/Feeling-stressed-Then-go-mow-the-lawn-claims-research.html* (accessed August 9, 2011).

9. Ontario Ministry of Agriculture, Food and Rural Affairs, "Climate Zones and Planting Dates for Vegetables in Ontario," *www.omafra.gov.on.ca/english/crops/facts/climzoneveg.htm*.

10. National Climate Data and Information Archive, "Canadian Climate Normals or Averages 1971–2000," *www.climate.weatheroffice.gc.ca/climate_normals/index_e.html*.

11. Sunrise Sunset, "Custom Sunrise Sunset Calendar," *www.sunrisesunset.com*.

June

1. Brian Naylor, "Osprey," 168–69, in M.D. Cadman, D.A. Sutherland, G.G. Beck, D. Lepage, and A.R. Couturier, eds, *Atlas of the Breeding Birds of Ontario, 2001–2005*.

2. Kathy E. Jones and Steven T.A. Timmermans, "Common Loon," 140–41, in M.D. Cadman, D.A. Sutherland, G.G. Beck, D. Lepage, and A.R. Couturier, eds, *Atlas of the Breeding Birds of Ontario, 2001–2005*.

3. Peter Blancher, Michael D. Cadman, Bruce A. Pond, Andrew R. Couturier, Erica H. Dunn, Charles M. Francis, and Robert S. Rempel, "Changes in Bird Distributions Between Atlases," 42, in M.D. Cadman, D.A. Sutherland, G.G. Beck, D. Lepage, and A.R. Couturier, eds, *Atlas of the Breeding Birds of Ontario, 2001–2005*.

4. Cecily Ross, "Songs of the Bobolink," *ON Nature*, Summer 2010, 25.

5. The Canadian Eastern Massasauga Rattlesnake Recovery Team, *Eastern Massasauga Rattlesnake Stewardship Guide: A Resource and Field Guide for Living with Rattlesnakes in Ontario* (Environment Canada, Toronto Zoon, World Wildlife Fund, Ontario's Living Legacy), 20–21.

6. Ontario Ministry of Natural Resources, "Species at Risk: Great Lakes–St. Lawrence Forest," *www.mnr.gov.on.ca/en/Business/Species/2ColumnSubPage/276503.html*.

7. Glenda Clayton, "Troubling Times for Ontario's Turtles," CottageCountryNow, *www.cottage countrynow.ca/community/life/article/401213--troubling-times-for-ontario-apos-s-turtles.*

8. Mart R. Gross and Eric L. Charnov, 1980, "Alternative Male Life Histories in Bluegill Sunfish," *Proceedings of the National Academy of Science*, USA Vol. 77, No. 11, 6937–940.

9. Muskoka Watershed Council, "Climate Change and Adaptation in Muskoka" (April 2010), 3, *www.muskokaheritage.org/watershed/PDFs/ClimateChangeMuskoka_April2010.pdf.*

10. Wikipedia, "Tornadoes of 2006," *en.wikipedia.org/wiki/Tornadoes_of_2006.*

11. National Climate Data and Information Archive, "Canadian Climate Normals or Averages 1971–2000," *www.climate.weatheroffice.gc.ca/climate_normals/index_e.html.*

12. Sunrise Sunset, "Custom Sunrise Sunset Calendar," *www.sunrisesunset.com.*

July

1. Doug Bennett and Tim Tiner, *Wild City: A Guide to Nature in Urban Ontario, from Termites to Coyotes* (Toronto: McClelland & Stewart Ltd., 2004), 39–40.

2. Ontario Ministry of Natural Resources, "Status of Elk in Ontario," *www.mnr.gov.on.ca/en/Business/FW/2ColumnSubPage/279012.html.*

3. Ontario Ministry of Natural Resources, "Bear Wise," *www.mnr.gov.on.ca/en/Business/Bearwise.*

4. Ontario Ministry of Natural Resources, "Species at Risk: Great Lakes–St. Lawrence Forest," *www.mnr.gov.on.ca/en/Business/Species/2ColumnSubPage/276503.html.*

5. Service Ontario: e-laws, "Endangered Species Act, 2007 — Protection and Recovery of Species," *www.e-laws.gov.on.ca/html/statutes/english/elaws_statutes_07e06_e.htm#BK12.*

6. Ontario Nature, "Ontario's Reptile and Amphibian Atlas," *www.ontarionature.org/protect/species/herpetofaunal_atlas.php.*

7. Ontario Ministry of Natural Resources, "Kawartha Lakes Fish Die-offs in 2007: Summary Report," *www.mnr.gov.on.ca/stdprodconsume/groups/lr/@mnr/@letsfish/documents/document/241459.pdf.*

8. Paul Pratt, "Regional Lists of Ontario Odonata," *www.wincom.net/~prairie/odonata.html.*

9. OFAH Invading Species Awareness Program, "Zebra Mussel," *www.invadingspecies.com/indexen.cfm.*

10. _____, "Dog Strangling Vine."

11. _____, "Garlic Mustard."

12. _____, "Common Reed."

13. Personal communication, Mike Oldham, Natural Heritage Information Centre, email, October 16, 2011.

14. OFAH Invading Species Awareness Program, "Giant Hogweed," *www.invadingspecies.com/indexen.cfm*.

15. Janine McLeod, Education and Outreach Coordinator, Alderville First Nation Black Oak Savanna, email, September 22, 2011.

16. Ontario Ministry of Natural Resources, "Biodiversity: Progress on Fighting Invasive Species in Ontario," *www.mnr.gov.on.ca/en/Business/Biodiversity/2ColumnSubPage/STDPROD_068687.html*.

17. National Climate Data and Information Archive, "Canadian Climate Normals or Averages 1971–2000," *www.climate.weatheroffice.gc.ca/climate_normals/index_e.html*.

18. Environment Canada, "Lightning Safety," *ec.gc.ca/foudre-lightning/default.asp?lang=En&n=159F8282-1*.

19. Sunrise Sunset, "Custom Sunrise Sunset Calendar," *www.sunrisesunset.com*.

August

1. P.J. Wilson, S. Grewal, I.D. Lawford, J.N.M. Heal, A.G. Granacki, D. Pennock, J.B. Theberge, M.T. Theberge, D.R. Voigt, W. Waddell, R.E. Chambers, P.C. Paquet, G. Goulet, D. Cluff, and B.N. White. "DNA Profiles of the Eastern Canadian Wolf and the Red Wolf Provide Evidence for a Common Evolutionary History Independent of the Gray Wolf." *Canadian Journal of Zoology* 78: 2,156–166.

2. Mike Oldham, botanist, Natural Heritage Information Centre, email, October 16, 2011.

3. Don Comis, "Global Warming's High Carbon Dioxide Levels May Exacerbate Ragweed Allergies," Agricultural Research Service, United States Department of Agriculture, *www.ars.usda.gov/is/pr/2000/000815.htm*.

4. Agriculture and Agri-Food Canada, "Algae, Cyanobacteria and Water Quality," *www4.agr.gc.ca/AAFC-AAC/display-afficher.do?id=1189714026543&lang=eng*.

5. Agriculture and Agri-Food Canada, "Air Quality Ontario: Ground-level Ozone," *www.airqualityontario.com/science/pollutants/ozone.php*.

6. National Climate Data and Information Archive, "Canadian Climate Normals or Averages 1971–2000," *www.climate.weatheroffice.gc.ca/climate_normals/index_e.html*.

7. Sunrise Sunset, "Custom Sunrise Sunset Calendar," *www.sunrisesunset.com*.

September

1. Stephen T. Emlen, "Migratory Orientation in the Indigo Bunting," *The Auk* 84: 463–89. October, 1967.

2. Ontario Ministry of Natural Resources, "Walleye State of the Resource Report for Southern Region," *www.mnr.gov.on.ca/stdprodconsume/groups/lr/@mnr/@letsfish/documents/document/mnr_e001342.pdf*.

3. National Climate Data and Information Archive, "Canadian Climate Normals or Averages 1971–2000," *www.climate.weatheroffice.gc.ca/climate_normals/index_e.html*.

4. Sunrise Sunset, "Custom Sunrise Sunset Calendar," *www.sunrisesunset.com*.

October

1. Erica Nol, professor, biologist, Trent University, email, September 22, 2011.

2. Canadian Cooperative Wildlife Health Centre Resources, "Type E Botulism in Birds," *www.ccwhc.ca/wildlife_health_topics/botulism/botulisme_org.php*.

3. Ron Pittaway, "2011–2012 Winter Finch Forecast," *www.jeaniron.ca/2011/finchforecast.htm*.

4. Scott Camazine, *The Naturalist's Year: 24 Outdoor Explorations* (Toronto: John Wiley & Sons, Inc. 1987), 155.

5. National Climate Data and Information Archive, "Canadian Climate Normals or Averages 1971–2000," *www.climate.weatheroffice.gc.ca/climate_normals/index_e.html*.

6. Sunrise Sunset, "Custom Sunrise Sunset Calendar," *www.sunrisesunset.com*.

November

1. Bernd Heinrich, *Winter World: The Ingenuity of Animal Survival* (New York: Ecco, an Imprint of HarperCollins Publishers, 2003), 307.

2. The Science Behind Algonquin's Animals, "Painted Turtle," *www.sbaa.ca/projects.asp?cn=316*.

3. Christine Merlin, Robert J. Gegear, and Steven M. Reppert, "Antennal Circadian Clocks Coordinate Sun Compass Orientation in Migratory Monarch Butterflies," *Science*, September 25, 2009: Vol. 325, No. 5948, 1700–704.

4. Bernd Heinrich, *A Year in the Maine Woods* (Don Mills: Addison-Wesley Publishing Company, 1994), 179.

5. Andy Holdsworth, Cindy Hale, and Lee Frelich, "Invasive Terrestrial Animals: Earthworms," Minnesota Department of Natural Resources, *www.dnr.state.mn.us/invasives/terrestrialanimals/earthworms/index.html*.

6. National Climate Data and Information Archive, "Canadian Climate Normals or Averages 1971–2000," *www.climate.weatheroffice.gc.ca/climate_normals/index_e.html*.

7. Sunrise Sunset, "Custom Sunrise Sunset Calendar," *www.sunrisesunset.com*.

December

1. Audubon, "Christmas Bird Counts, Historical Results, 2009–10," *birds.audubon.org/historical-results*.

2. Don Sutherland, zoologist, Natural Heritage Information Centre, email, May 13, 2011.

3.   Bob Bergmann, Regional Fisheries biologist, Ministry of Natural Resources, email, October 20, 2011.

4.   Explore Ontario's Biodiversity, "American Eel," *www.rom.on.ca/ontario/index.php*.

5.   Ontario Ministry of Natural Resources, "Forest Health Management — Insects and Diseases and Invasive Species," *www.mnr.gov.on.ca/en/Business/Forests/2ColumnSubPage/STEL02_166920.html*.

6.   Doug Sadler, *Winter: A Natural History* (Peterborough: Broadview Press, 1990), 179.

7.   Peter Adams and Colin Taylor, eds, *Peterborough and the Kawarthas*, 3rd Edition (Peterborough: Trent University Geography Department, 2009), 105–06.

8.   The Weather Doctor, "Lake-Effect Snow Climatology in the Great Lakes Region," *www.islandnet.com/~see/weather/doctor.htm*.

9.   National Climate Data and Information Archive, "Canadian Climate Normals or Averages 1971–2000," *www.climate.weatheroffice.gc.ca/climate_normals/index_e.html*.

10. Sunrise Sunset, "Custom Sunrise Sunset Calendar," *www.sunrisesunset.com*.

BIBLIOGRAPHY

Adams, P., and Colin Taylor, eds. *Peterborough and the Kawarthas*. Peterborough, ON: Heritage Publications, 1985.

Adams, Peter, and Colin Taylor, eds. *Peterborough and the Kawarthas* 3rd Edition. Peterborough, ON: Trent University Geography Department, 2009.

Agriculture and Agri-Food Canada. "Air Quality Ontario: Ground-level Ozone." *www.airquality ontario.com/science/pollutants/ozone.php*.

_____. "Algae, Cyanobacteria and Water Quality." *www4.agr.gc.ca/AAFC-AAC/display-afficher. do?id=1189714026543&lang=eng*.

Ahas, R., Mark D. Schwartz, and A. Aasa. "Onset of Spring Starting Earlier Across the Northern Hemisphere." *Global Change Biology* 12: 343–51.

Alsheimer, Charles J. *Whitetail: Behaviour Through the Seasons*. Iola, WI: Krause Publications, 1996.

Barron, George. *Mushrooms of Ontario & Eastern Canada*. Edmonton: Lone Pine Publishing, 1999.

Bates, J. *A Northwoods Companion: Fall & Winter*. Mercer, WI: Manitowish River Press, 1997.

Bennett, Doug, and Tim Tiner. *Up North Again*. Toronto: McClelland & Stewart, 1997.

_____. *Up North Daybook 1995*. Markham, ON: Reed Books, 1995.

_____. *Wild City: A Guide to Nature in Urban Ontario, from Termites to Coyotes*. Toronto: McClelland & Stewart Ltd., 2004.

Borland, Hal. *Beyond Your Doorstep: A Handbook to the Country*. Guilford, CT: The Lyons Press, 2003.

_____. *Book of Days*. New York: Alfred A. Knopf, 1976.

_____. *Twelve Moons of the Year*. New York: Alfred A. Knopf, 1979.

Brown, Lauren. *Grasses: An Identification Guide*. New York: Houghton Mifflin, 1979.

Burke, Peter S., Colin D. Jones, Jennifer M. Line, Michael J. Oldham, and Peter J. Sorrill. *1998 Peterborough County Natural History Summary*. Peterborough, ON: Peterborough Field Naturalists, Natural Heritage Information Centre, and Trent University, 1999.

Cadman, M., P. Eagles, and F. Helleiner. *Atlas of the Breeding Birds of Ontario*. Waterloo, ON: University of Waterloo Press, 1987.

Cadman, M.D., D.A. Sutherland, G.G. Beck, D. Lepage, and A.R. Couturier, eds. *Atlas of the Breeding Birds of Ontario, 2001–2005*. Toronto: Bird Studies Canada, Environment Canada, Ontario Field Ornithologists, Ontario Ministry of Natural Resources, and Ontario Nature, 2007.

Camazine, Scott. *The Naturalist's Year: 24 Outdoor Explorations.* Toronto: John Wiley & Sons, Inc. 1987.

Canadian Cooperative Wildlife Health Centre Resources. "Type E Botulism in Birds." *www. ccwhc.ca/wildlife_health_topics/botulism/botulisme_org.php.*

Canadian Eastern Massasauga Rattlesnake Recovery Team (The). *Eastern Massasauga Rattlesnake Stewardship Guide: A Resource and Field Guide for Living with Rattlesnakes in Ontario.* Environment Canada, Toronto Zoo, World Wildlife Fund, Ontario's Living Legacy.

Carpentier, Geoff. *The Mammals of Peterborough County.* Peterborough, Ontario: Peterborough Field Naturalists, 1987.

Chambers, Brenda, Karen Legasy, and Cathy V. Bentley. *Forest Plants of Central Ontario.* Edmonton: Lone Pine Publishing, 1996.

Chu, Miyoko. *Songbird Journeys: Four Seasons in the Lives of Migratory Birds.* New York: Walker & Company, 2007.

Clayton, Glenda. "Troubling Times for Ontario's Turtles." *CottageCountryNow. www.cottage countrynow.ca/community/life/article/401213--troubling-times-for-ontario-apos-s-turtles.*

Comis, Don. "Global Warming's High Carbon Dioxide Levels May Exacerbate Ragweed Allergies." Agricultural Research Service, United States Department of Agriculture. *www.ars.usda.gov/is/pr/2000/000815.htm.*

Connor, Jack. *The Complete Birder.* Boston: Houghton Mifflin Co., 1988.

Conservation Ontario. "Your Guide to the Conservation Areas in Ontario." *www.ontarioconserv ationareas.ca.*

Dukenvironment Magazine Fall 2006. "Global Warming May Increase Poison Ivy's Itch." *www. nicholas.duke.edu/DEarchives/f06/log-ivy.html.*

Dunkle, Sidney. *Dragonflies Through Binoculars: A Field Guide to Dragonflies of North America.* New York: Oxford University Press.

Elliott, Lang, and Wil Hershberger. *The Songs of Insects.* Boston: Houghton Mifflin Company, 2007.

Emlen, Stephen T. "Migratory Orientation in the Indigo Bunting." *The Auk,* 84.

Environment Canada. "Lightning Safety." *ec.gc.ca/foudre-lightning/default.asp?lang=En&n= 159F8282-1.*

Explore Ontario's Biodiversity. "American Eel." *www.rom.on.ca/ontario/index.php.*

Farrar, John Laird. *Trees in Canada.* Ottawa: Fitzhenry & Whiteside Ltd and the Canadian Forest Service, 1995.

Follette, Robin. "Black Fly Info." Maine Nature Notes. *mainenature.org/black-fly-info.*

Foote, Jennifer R., and others. "Black-capped Chickadee (*Poecile atricapillus*)." The Birds of North America Online. *bna.birds.cornell.edu/bna/species/039.*

George, Catherine. "Fall Foliage Guide." *Toronto Star,* September 12, 2009. *www.thestar.com/ travel/ontariooutings/article/693265.*

Gross, Mart R., and Eric L. Charnov. "Alternative Male Life Histories in Bluegill Sunfish." *Proceedings of the National Academy of Science*, U.S.A., Vol. 77, No. 11.

Heidorn, Keith C. *And Now … the Weather: With the Weather Doctor*. Calgary: Fifth House, 2005.

Heinrich, Bernd. *A Year in the Maine Woods*. Don Mills, ON: Addison-Wesley, 1994.

_____. *Summer World: A Season of Bounty*. New York: Ecco, an Imprint of HarperCollins Publishers, 2009.

_____. *Winter World: The Ingenuity of Animal Survival*. New York: Ecco, an imprint of Harper-Collins Publishers, 2003.

Holdsworth, Andy, Cindy Hale, and Lee Frelich. "Invasive Terrestrial Animals: Earthworms." Minnesota Department of Natural Resources. *www.dnr.state.mn.us/invasives/terrestrial animals/earthworms/index.html*.

Holm, Erling, Nick Mandrak, and Mary Burridge. *The ROM Field Guide to Freshwater Fishes of Ontario*. Toronto: Royal Ontario Museum, 2009.

Jauss, Anne Marie. *Discovering Nature the Year Round*. New York: E.P. Dutton & Co., 1955.

Johnson, Lorraine. "Save the Bees: How to Attract Pollinators to Your Yard." *Canadian Gardening*. *www.canadiangardening.com/how-to/wildlife*.

Jones, Colin, Andrea Kingsley, Peter Burke, and Matt Holder. *Field Guide to the Dragonflies and Damselflies of Algonquin Provincial Park and the Surrounding Area*. Whitney, ON: The Friends of Algonquin Park, 2008.

Kaufman, Kenn. *Field Guide to Advanced Birding: Understanding What You See and Hear*. Boston: Houghton Mifflin Harcourt, 2011.

Lam, Ed. *Damselflies of the Northeast: A Guide to the Species of Eastern Canada and the Northeastern United States*. Forest Hills, NY: Biodiversity Books, 2004.

Land Between (The). "Nature." *www.thelandbetween.ca/naturalheritage.asp*.

Lawrence, Steve. "Climate Change Putting Lake Trout at Risk." Haliburton Highlands Outdoors Association. *www.hhoa.on.ca/education.html*.

Layberry, Ross A., Peter W. Hall, and J. Donald Lafontaine. *The Butterflies of Canada*. Toronto: University of Toronto Press, 1998.

Leopold, A., and S.E. Jones. "A Phenological Record for Sauk and Dane Counties, Wisconsin, 1935–45" *Ecological Monographs*. Vol. 17, No. 1, (1947): 81–122.

Leopold, Aldo. *A Sand County Almanac*. New York: Ballantine Books, 1970.

Logier, E.B. *The Snakes of Ontario*. Toronto: University of Toronto Press, 1958.

Lorch, Jeffrey M., and others. "Experimental Infection of Bats with *Geomyces destructans* Causes White-nose Syndrome." *Nature* (2011) doi: 10.1038/nature10590. Published online October 26, 2011. *www.nature.com/nature/journal/vaop/ncurrent/full/nature10590.html*.

Marshall, Stephen A. *Insects: Their Natural History and Diversity*. Richmond Hill, ON: Firefly Books Ltd., 2006.

Mechler, Gary, Wil Tirion, and Mark Chartrand. *National Audubon Society Pocket Guide: Constellations*. New York: Alfred A. Knopf, Inc., 1995.

Merlin, Christine, Robert J. Gegear, and Steven M. Reppert. "Antennal Circadian Clocks Coordinate Sun Compass Orientation in Migratory Monarch Butterflies." *Science*, September 25, 2009: Vol. 325, No. 5, 948.

Metcalfe-Smith, Janice, Alistair MacKenzie, Ian Carmichael, and Daryl McGoldrick. *Photo Field Guide to the Freshwater Mussels of Ontario*. St. Thomas, ON: St. Thomas Field Naturalist Club Incorporated, 2005.

Monkman, Drew. *Nature's Year in the Kawarthas: A Guide to the Unfolding Seasons*. Toronto: Natural Heritage, 2002.

Muskoka Watershed Council. "Climate Change and Adaptation in Muskoka." *www.muskoka heritage.org/watershed/PDFs/ClimateChangeMuskoka_April2010.pdf*.

National Climate Data and Information Archive. "Canadian Climate Normals or Averages 1971–2000." *www.climate.weatheroffice.gc.ca/climate_normals/index_e.html*.

National Public Radio. "Scientists Say Bat Disease Likely to Spread." Interview with Brock Fenton. *www.npr.org/templates/story/story.php?storyId=112035629*.

Nelson, Roger. *A Dales Naturalist*. Skipton, UK: Dalesman Publishing Co., 1993.

Newcomb, Lawrence. *Newcomb's Wildflower Guide*. Toronto: Little, Brown and Company, 1977.

Nikula, Blair, Jackie Sones, Donald Stokes, and Lillian Stokes. *Stokes Beginner's Guide to Dragonflies and Damselflies*. Boston: Little, Brown and Company, 2002.

OFAH Invading Species Awareness Program. "Common Reed." *www.invadingspecies.com/indexen. cfm*.

_____. "Dog Strangling Vine."

_____. "Garlic Mustard."

_____. "Giant Hogweed."

_____. "Zebra Mussel."

Ontario Field Ornithologists. "Site Guides and Hotspots." *www.ofo.ca/hotspots/siteguides.php*.

Ontario Ministry of Agriculture, Food and Rural Affairs. "Climate Zones and Planting Dates for Vegetables in Ontario." *www.omafra.gov.on.ca/english/crops/facts/climzoneveg.htm*.

Ontario Ministry of Natural Resources. "Bear Wise." *www.mnr.gov.on.ca/en/Business/Bearwise*.

_____. "Biodiversity: Progress on Fighting Invasive Species in Ontario." *www.mnr.gov.on.ca/en/ Business/Biodiversity/2ColumnSubPage/STDPROD_068687.html*.

_____. "Forest Health Management — Insects and Diseases and, Invasive Species." *www.mnr.gov. on.ca/en/Business/Forests/2ColumnSubPage/STEL02_166920.html*.

_____. "Kawartha Lakes Fish Die-offs in 2007: Summary Report." *www.mnr.gov.on.ca/stdprod consume/groups/lr/@mnr/@letsfish/documents/document/241459.pdf*.

_____. "Species at Risk: Great Lakes–St. Lawrence Forest." *www.mnr.gov.on.ca/en/Business/ Species/2ColumnSubPage/276503.html.*

_____. "Status of Elk in Ontario." *www.mnr.gov.on.ca/en/Business/FW/2ColumnSubPage/279012. html.*

_____. "Walleye State of the Resource Report for Southern Region." *www.mnr.gov.on.ca/stdprod consume/groups/lr/@mnr/@letsfish/documents/document/mnr_e001342.pdf.*

Ontario Nature. "Ontario's Reptile and Amphibian Atlas." *www.ontarionature.org/protect/species/ herpetofaunal_atlas.php.*

_____. "Nature Reserves." *www.ontarionature.org/protect/habitat/nature_reserves.php.*

Ontario Parks. "Park Locator." *www.ontarioparks.com/english/locator.html.*

Ontario Tourism Marketing Partnership. "Places We Love: Great Fall Drives." *www.ontariotravel. net/TCISSegmentsWeb/gc/FD?language=en.*

Parks Canada. "National Parks List." *www.pc.gc.ca/progs/np-pn/recherche-search_e.asp?p=1.*

Pelletier, Georges. *Insectes Chanteurs du Québec.* L'Acadie, QC: Editions Broquet, 1995.

Peterborough Field Naturalists. *Kawarthas Nature.* Erin, ON: Boston Mills Press, 1992.

Philipp, D.P., and M.R. Gross. "Genetic Evidence for Cuckoldry in Bluegill *Lepomis marcrochirus.*" *Molecular Ecology* (3) (1994): 563–69.

Pittaway, Ron. "2011–2012 Winter Finch Forecast." *www.jeaniron.ca/2011/finchforecast.htm.*

Pratt, Paul. "Regional Lists of Ontario Odonata." *www.wincom.net/~prairie/odonata.html.*

Project FeederWatch 2010–2011. "Top 25 Feeder Birds 2010–2011 Ontario." *www.bsc-eoc.org/ volunteer/pfw/index.jsp?lang=EN&targetpg=index.*

Regional Municipality of Halton: Public Health. "West Nile Virus and Mosquitoes." *www.halton. ca/cms/one.aspx?portalId=8310&pageId=10023.*

Rey, H.A. *The Stars.* Boston: Houghton Mifflin, 1980.

Ross, Cecily. "Songs of the Bobolink." *ON Nature,* Summer 2010, 25.

Runtz, Michael W.P. *The Explorer's Guide to Algonquin Park.* Toronto: Stoddart Publishing, 1993.

Sadler, Doug. *Our Heritage of Birds.* Peterborough, ON: Peterborough Field Naturalists, 1983.

_____. *Winter: A Natural History.* Peterborough, ON: Broadview Press, 1990.

The Science Behind Algonquin's Animals. "Painted Turtle." *www.sbaa.ca/projects.asp?cn=316.*

Scott, W.B. *Freshwater Fishes of Eastern Canada.* Toronto: University of Toronto Press, 1967.

Seidl, Amy. *Early Spring: An Ecologist and Her Children Wake to a Warming World.* Boston: Beacon Press, 2009.

Serrao, John. *Nature's Events.* Harrisburg, PA: Stackpole Books, 1992.

Service Ontario: e-laws. "Endangered Species Act, 2007 — Protection and Recovery of Species." *www.elaws.gov.on.ca/html/statutes/english/elaws_statutes_07e06_e.htm#BK12.*

Sheldon, Ian. *Animal Tracks of Ontario*. Edmonton: Lone Pine Publishing, 1997.

Sibley, David Allen. *The Sibley Field Guide to Birds of Eastern North America*. New York: Alfred A. Knopf, 2003.

Stokes, Donald. *A Guide to Nature in Winter*. Toronto: Little, Brown and Company, 1976.

_____. *A Guide to Observing Insect Lives*. Toronto: Little, Brown and Company, 1983.

Stokes, Donald, and Lillian Stokes. *A Guide to Bird Behaviour: Volume III*. Toronto: Little, Brown and Company, 1989.

Strickland, Dan, and Russ Rutter. *The Best of the Raven*. Whitney, ON: The Friends of Algonquin Park, 1993.

"Summary of the 110th Christmas Bird Count." *American Birds 2009–2010*. Vol. 64.

Sunrise Sunset. "Custom Sunrise Sunset Calendar." *www.sunrisesunset.com*.

Taggart, Ken (compiler). "Kawartha Lakes Northern Pike Invasion Project (2009–Present)." *Muskies Canada*. www.muskiescanada.ca/club_information/initiatives.php.

Teale, Edwin Way. *A Walk Through the Year*. New York: Dodd, Mead, and Co., 1987.

Theberge, John B., ed. *Legacy: The Natural History of Ontario*. Toronto: McClelland & Stewart, 1989.

Union of Concerned Scientists, "Climate Change and Ontario." *www.ucsusa.org/greatlakes/glregionont.html*.

U.S. Environmental Protection Agency. "Natural Processes in the Great Lakes." *www.epa.gov/greatlakes/atlas/glat-ch2.html*.

Victory Seeds. "Average First and Last Frost Dates for Ontario, Canada." *www.victoryseeds.com/frost/ontario.html*.

Wake, Winifred (Cairns), John Cartwright, Anne Champagne, Kathy Parker, and Martin Parker, eds. *A Nature Guide to Ontario: Federation of Ontario Naturalists*. Toronto: University of Toronto Press, 1997.

Walton, Richard, and Robert Lawson. *Birding By Ear: Eastern/Central* (Peterson Series). Boston: Houghton Mifflin, 1990.

The Weather Doctor. "Lake-Effect Snow Climatology in the Great Lakes Region." *www.islandnet.com/~see/weather/doctor.htm*.

Weber, Larry. *Backyard Almanac*. Duluth, MN: Pfeifer-Hamilton Publishers, 1996.

Whelan, Peter. "Birds" (column) *Globe and Mail*, 1985–99.

Wikipedia. "Tornadoes of 2006." *en.wikipedia.org/wiki/Tornadoes_of_2006*.

Wilson, P.J., S. Grewal, I.D. Lawford, J.N.M. Heal, A.G. Granacki, D. Pennock, J.B. Theberge, M.T. Theberge, D.R. Voigt, W. Waddell, R.E. Chambers, P.C. Paquet, G. Goulet, D. Cluff, and B.N. White. "DNA Profiles of the Eastern Canadian Wolf and the Red Wolf Provide Evidence for a Common Evolutionary History Independent of the Gray Wolf." *Canadian Journal of Zoology*: 78.

List of Websites

As you can well imagine, the number of nature sites on the web is staggering. However, these are some of the sites that I find most useful.

General

Animal Diversity Web (*animaldiversity.ummz.umich.edu/site/index.html*) This is an online database of the natural history of thousands of species from the different classes of the animal kingdom from invertebrates to mammals. The species accounts even include movies of specimens and recordings of sounds.

Boreal Forest (*www.borealforest.org/edresc.htm*) Despite the site's name, nearly all of the species descriptions apply equally well to central and eastern Ontario. I find the site most useful for the information about trees, flowers, ferns, sedges, grasses, mosses, and lichens.

eNature (*www.enature.com/home/indexNew.asp*) This site is possibly the web's premier destination for information about the wild animals and plants of North America. The site's core content is the same data set used to create the printed Audubon Field Guides. It also includes excellent guides to the night sky.

Encyclopedia of Life (EOL) (*www.eol.org*) The EOL is no less than a free online collaborative encyclopedia and database for every one of the 1.8 million species that are named and known on this planet.

Explore Ontario's Biodiversity (*www.rom.on.ca/ontario/index.php*) This Royal Ontario Museum site allows you to create your own field guide to animals of your local area. You can also obtain lists of Ontario species at risk.

The Land Between (*www.thelandbetween.ca*) A large slice of central Ontario, namely the transition zone that lies between the Canadian Shield and the St. Lawrence Lowlands stretching from Georgian Bay to Kingston. You will find a plethora of information ranging from the area's natural and cultural features to conservation initiatives.

Natural Heritage Information Centre (*nhic.mnr.gov.on.ca*) The NHIC compiles, maintains, and distributes information on natural species, plant communities, and spaces of conservation concern in Ontario. You can use the "Biodiversity Explorer" for querying records of biodiversity information within Ontario.

Ontario Nature (*www.ontarionature.org/index.php*) A charitable organization that protects wild species and wild spaces. The website contains a wealth of resources to help you get involved in conservation projects. There are also excellent downloadable resources such as "A Citizen's Toolkit for Nature Conservation."

BIRDS

Atlas of the Breeding Birds of Ontario (*www.birdsontario.org/atlas/index.jsp*) By selecting the "Data and Maps" tab, you can find maps for both the breeding range and the breeding abundance for all of Ontario's breeding birds. You can also compare the second atlas (2001–05) to the first (1981–85).

BirdJam (*www.birdjam.com/learn.php*) This is a great site for starting to learn bird sounds and songs.

The Birds of North America Online (*bna.birds.cornell.edu/bna*) This site provides comprehensive life histories for each of the 716+ species of birds breeding in the United States and Canada. There are also image and video galleries showing plumages, behaviours, habitats, nests, eggs, and more. You will need to purchase a subscription.

Cornell Lab of Ornithology (*www.birds.cornell.edu/Page.aspx?pid=1478*) Cornell is a world leader in the study, appreciation, and conservation of birds. Click on "All About Birds" for an online bird guide that includes vocalizations.

eBird Canada (*ebird.org/content/canada*) Offering a real-time, online checklist program, at Ebird you will find rich data on bird abundance and distribution, including seasonal bar charts for every county in Ontario.

Ontario Field Ornithologists (*www.ofo.ca*) Here you'll find photos of recent bird sightings, field trip information, and much more. You will also find ONTBIRDS, OFO's electronic mailing LISTSERV, which notifies birders of interesting Ontario bird sightings.

MAMMALS

Guide to New York State Mammals (*www.nyfalls.com/wildlife/Wildlife-mammals.html*) Central and eastern Ontario and New York State share pretty much the same mammals. You will find extensive information here on species from all of the mammal families. Audio is included with the vocal mammals.

REPTILES AND AMPHIBIANS

Adopt-A-Pond (*www.torontozoo.com/AdoptAPond*) This site provides information resources and educational opportunities to conserve and create wetland habitats. Click on the species guides to learn about Ontario's amphibians and reptiles.

NatureWatch (*www.naturewatch.ca*) NatureWatch offers a series of ecological monitoring programs that encourage you to become a citizen scientist. Click on "FrogWatch" for information on frogs and to learn their calls.

FISH

Fishes of Canada (*www.aquatic.uoguelph.ca/fish/fish.htm*) Here you'll find a treasure trove of information on Canada's fishes, including descriptions, distribution, ecology, and reproduction. You'll also find all of the non-game species.

Go Fish in Ontario (*www.gofishinontario.com*) This site provides a wealth of information on all aspects of sport fishing in Ontario, including lots of facts on the fish themselves.

Let's Fish Ontario (*www.mnr.gov.on.ca/en/Business/LetsFish/index.html*) This Ministry of Natural Resources site offers information on fishing licences, fishing regulations, fish facts, and much more.

INVERTEBRATES

BugGuide (*bugguide.net/node/view/15740*) BugGuide is an online community of naturalists who enjoy learning about and sharing photographs and observations about insects, spiders, and other related invertebrates. Use the clickable guide to identify most any insect-like creature you might come across.

Odonata Central (*www.odonatacentral.org*) Odonata Central makes available what is known about the distribution, biodiversity, and identification of Odonata (dragonflies and damselflies). Start with the checklist feature to get a list of those species found in Ontario. You can then go on to browse the photographs.

Toronto Entomologists' Association (*www.ontarioinsects.org*) A great site for anyone interested in learning more about Ontario's insects.

PLANTS

Andy's Northern Ontario Wildflowers (*www.ontariowildflower.com*) A richly illustrated site to the wildflowers of the Sudbury area, almost all of which are also found in central and eastern Ontario.

Field Botanists of Ontario (*www.trentu.ca/org/fbo*) The FBO is open to anyone with an interest in botany and its conservation in the province of Ontario.

Ontario Trees and Shrubs (*ontariotrees.com/index.php*) An online "field guide" to Ontario's trees and shrubs. Search by name, habitat, colour, leaf characteristics, or just about any other criteria.

Ontario Wildflowers (*ontariowildflowers.com/index.php*) The same as above, but this time for wildflowers.

FUNGI

Mushroom Expert (*www.mushroomexpert.com*) This site provides everything you'll need to study and identify mushrooms.

WEATHER AND CLIMATE

Weather Office (*www.weatheroffice.gc.ca*) This Environment Canada site not only provides current weather conditions and forecasts but also historical weather.

Center for Climate and Energy Solutions (C2ES) (*www.pewclimate.org*) Anyone with a love for the natural world and the changing seasons has to be concerned about the threat of climate change. The C2ES Center was named the world's top environmental think tank in 2011. This site covers all aspects of the global debate on climate change, including the basic science involved, a Kids Corner, and how to talk to climate skeptics.

ASTRONOMY

Earth Sky (*earthsky.org/tonight*) This site will show you the most interesting and noteworthy planets, stars, and constellations that are visible on any given night.

PHENOLOGY

Phenology concerns the observation and study of cyclical biological phenomena such as emerging leaves, fruiting trees, and migration in relation to climatic conditions.

Coffrin Center for Biodiversity (*www.uwgb.edu/biodiversity/phenology/index.htm*) Here, you can view phenological records — weather, plant, bird events, etc. — for almost every day of the year in Green Bay, Wisconsin. Green Bay is at the same latitude as southern central Ontario and has generally the same mix of plants and animals.

Our Changing Seasons (*www.drewmonkman.com*) My website provides a record of interesting nature sightings occurring in Peterborough County as well as an archive of my *Peterborough Examiner* nature columns.

Step Outside (*www.r4r.ca/en/step-outside/nature-guides*) Published twice a month, Step Outside is a compilation of seasonal happenings in Ontario. You'll find details, photographs, and interesting links regarding flora, fauna, the night sky, and climate events. This site is of particular interest to teachers.

This Week At Hilton Pond (*www.hiltonpond.org/ThisWeekMain.html*) This website is mostly devoted to weekly updates about phenology at Hilton Pond Center in York, South Carolina. Surprisingly, many of these accounts of the plants and animals apply equally well to central and eastern Ontario.

WILDLIFE IN DISTRESS

Kawartha Turtle Trauma Centre (*www.kawarthaturtle.org*) The Kawartha Turtle Trauma Centre is a non-profit, registered charity that operates a hospital for injured wild turtles. It is located in Peterborough, Ontario. Once healed, these turtles are released back into their natural habitat. (Tel: 705-741-5000)

Toronto Wildlife Centre (*www.torontowildlifecentre.com*) A registered charity that rescues wildlife in distress and provides them medical care and rehabilitation. (Tel: 416-631-9942)

INDEX

(N.B. Illustrations are indicated by italics.)

OF RELATED INTEREST

Nature in the Kawarthas
Compiled by
Gordon Berry, John Bottomley,
and Rebecca Zeran
for the Peterborough Field Naturalists
978-1459701151 | $32.99

This book is a valuable asset that presents a wealth of information on the birds, mammals, insects, flowers, reptiles, and amphibians of this special area. It describes where to visit to explore the natural wonders of this amazing area and its treasure of wild biodiversity. It is a true layman's guide to nature in the Kawarthas region of Ontario.

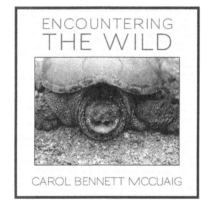

Encountering the Wild
Carol Bennett McCuaig
978-1554888580 | $25.00

Poison Acres, 250 acres of wilderness in Renfrew County, Ontario, dedicated to the preservation of natural habitat, has been home to the author for many years. Her keen powers of observation coupled with her insights into wildlife behaviour and her evocative writing style have produced this captivating collection of stories that will appeal to country lovers.

Available at your favourite bookseller.

 **DUNDURN**
www.dundurn.com

What did you think of this book?
Visit www.dundurn.com for reviews, videos, updates, and more!